Cultural Heritage and Archaeological Issues in Materials Science

MATERIALS RESEARCH SOCIETY
SYMPOSIUM PROCEEDINGS VOLUME 1374

Cultural Heritage and Archaeological Issues in Materials Science

EDITORS

Jose Luis Ruvalcaba Sil
Universidad Nacional Autonoma de Mexico
Mexico City, Mexico

Javier Reyes Trujeque
Universidad Autonoma de Campeche
Campeche, Mexico

Adrian Velazquez Castro
Instituto Nacional de Antropologia e Historia
Mexico City, Mexico

Manuel Espinosa Pesqueira
Instituto Nacional de Investigaciones Nucleares
Ocoyoacac, Mexico

Materials Research Society
Warrendale, Pennsylvania

CAMBRIDGE
UNIVERSITY PRESS

Shaftesbury Road, Cambridge CB2 8EA, United Kingdom

One Liberty Plaza, 20th Floor, New York, NY 10006, USA

477 Williamstown Road, Port Melbourne, VIC 3207, Australia

314–321, 3rd Floor, Plot 3, Splendor Forum, Jasola District Centre, New Delhi – 110025, India

103 Penang Road, #05–06/07, Visioncrest Commercial, Singapore 238467

Cambridge University Press is part of Cambridge University Press & Assessment, a department of the University of Cambridge.

We share the University's mission to contribute to society through the pursuit of education, learning and research at the highest international levels of excellence.

www.cambridge.org
Information on this title: www.cambridge.org/9781605113517

Materials Research Society
506 Keystone Drive, Warrendale, PA 15086, USA
http://www.mrs.org

First published 2012

CODEN: MRSPDH

A catalogue record for this publication is available from the British Library

ISBN   978-1-605-11351-7   Hardback

CONTENTS

ARCHAEOLOGICAL SCIENCE

CONSERVATION STUDIES

BIOMATERIALS TOPICS

METHODOLOGIES AND INSTRUMENTATION

PREFACE

The aim of this book is the dissemination of interdisciplinary investigations that involve scientific analysis of materials related to problems of cultural heritage, ranging from studies on identification and use of materials, ancient technologies, dating, and deterioration to restoration and conservation.

In Mexico, during the past 15 years, scientific investigations focusing on questions concerning material culture have involved interdisciplinary approaches and included the work of art historians, museum curators, conservators, archaeologists, paleontologists, biologists and even contemporary artists, all interested in getting involved more and more in the study of cultural materials. In particular in developing countries, research teams engaged in new trends of archaeometric studies and science in conservation are working on several fronts; one is the constant struggle to achieve a fully interdisciplinary approach, which cannot be taken just as a routine analysis. Moreover, in order to complete successfully these kinds of studies and to achieve not only academic success but to create a social impact that directly contributes to the preservation and the enrichment of the culture of each country, ingenuity, commitment, teamwork, development of new methodologies and equipments, economic resources and critical thinking are required. On the other hand, the interaction of different disciplines generates a new language that facilitates communication between scientists and specialists that have no detailed knowledge of the material and natural sciences.

These trends led to the creation of the symposium "Electron Microscopy in Archaeology and Art" in the framework of the 14th International Congress on Electron Microscopy (14th ICEM 1998) held in Cancun, Quintana Roo, Mexico. In this first meeting, a variety of investigations were presented that focused on the scientific analysis of pre-Hispanic and XVI century pigments, deterioration of pre-Hispanic monuments, consolidating materials in Mayan masks and more. This event presented the opportunity to show how instrumental scientific analysis was the answer to questions generated from the archaeological data and materials.

By the year 2000, the 32nd International Symposium on Archaeometry was organized in Mexico City by the Universidad Nacional Autonoma de Mexico (UNAM), the Instituto National de Investigaciones Nucleares (ININ) and the Instituto Nacional de Antropologia e Historia (INAH). For the first time a Latin-American country hosted such an event, bringing together an important number of scholars from around the world, which presented their research in ten different symposia. This conference aimed at linking the international archaeometry scene with the archaeological heritage of the American continent. The experience gained through the above mentioned events and the efforts of scholars in Mexico, helped to create the symposium called "Archaeological and Arts Issues in Materials Science", which was celebrated in the framework of the International Materials Research Congress of the Mexican Society of Materials from 2000 to 2010. This forum has enjoyed steady growth due to the continuous effort of the chairpersons, and the sustained interest of the participants that each year honored the symposium with their presence; today the event is a point of reference for materials science in archaeology, not only in Mexico but also internationally. The most relevant investigations of the symposia of 2003 through 2006 were selected by an academic committee and published in a series of books called "La Ciencia de Materiales y su Impacto en la Arqueología" volumes I, II, III & IV. This was possible thanks the sponsorship of the Mexican Academy of Materials (now named Mexican Society of Materials), and the financial support of various Mexican universities.

Reflecting the increased importance of materials science and archaeometry in Mexico, in 2009 the joint meeting of the symposium "Archaeological and Arts Issues in Materials Science" and the "2nd Latin-American Symposium on Physical and Chemical Methods in Archaeology, Art and Cultural Heritage Conservation" (LASMAC2009) was included in the XVIII International Material Research Congress. The objective of this symposium was to present and discuss the most recent Latin-American research on cultural heritage and the historical past, using a wide spectrum of techniques and scientific methodologies available in this geographic area. In this forum scientists and specialists in cultural heritage and archaeology from countries including Argentina, Brazil, Colombia, Chile, Cuba, Mexico, Spain, France and United States presented "the state of the art" in this subject.

The development of archaeometric research in Mexico during the past decade has closed the gap that separated it from the high standards of research performed in developed countries with more resources. Due to the constant evolution of this research and the increase of scientific topics included in the "Archaeological and Arts Issues in Materials Science" symposium, in 2011 it changed its name to "Cultural Heritage and Archaeological Issues in Materials Science (CHARIMSc)". Under this new name the symposium drew physicists, chemists, engineers, conservators, archaeologists, art historians, architects, restorers and other specialists involved in the scientific study of cultural heritage and archaeological and historical collections, and together they showed the impressive results of their scientific and interdisciplinary research.

Over 90 papers were presented during this symposium, including invited lectures from France, Italy, Israel, Thailand and Mexico, as well as oral and poster sessions of participants from Spain, Mexico, Italy, Russia, Poland, United States and Australia. This represents a substantial increment in the quality and the amount of papers and the number of participants compared to previous symposia and is a clear sign of the intensive work carried out over the last few years.

This book of proceedings is published by the Materials Research Society and Cambridge University Press, and it is the first one to bear the title "Cultural Heritage and Archaeological Issues in Materials Science (CHARIMSc)". This publication presents selected contributions of the symposium in its 2011 edition, evaluated by an academic committee. Main topics such as Non-destructive Characterization of Cultural Heritage, Technical Studies in Art History, Archaeological Science, Conservation Studies, Biomaterials Research and Methodologies & Instrumentation show the new trends and directions being taken worldwide in the analysis of cultural and archaeological materials.

The most diverse techniques and scientific methodologies were presented including non-destructive methods, spectroscopic techniques, ion beam accelerator techniques, electron and optical microscopy, imaging techniques, experimental archaeology, all kinds of chemical methods, dating, biological methodologies, deterioration studies and conservation procedures.

The techniques mentioned above were applied to the study of European paintings by hyperspectral near-infrared imaging, inks in manuscripts, dyes and pigments from Europe, Israel and pre-Columbian Mexico, glass in pictorial European techniques, pre-Hispanic manufacturing technologies of stone and shells in Mesoamerica, technological studies of metals from the Maya area, as well as modern mural painting on cement and modern materials such as latex sculptures from contemporary art.

Papers on science in conservation are related to bio-degradation of monuments, evaluation of conservation of shells in pre-Hispanic collections, modified clays for restoration purposes and environmental studies, among others subjects.

Some new methodologies for the identification of dyes, the study of biological remains such as bone and plants (maize), as well as satellite radar prospection are also included in this book.

Finally the readers will realize that the application of scientific methods on cultural heritage and archaeological materials is making an important impact on society worldwide.

Jose Luis Ruvalcaba Sil
Instituto de Fisica, UNAM. Mexico
Manuel Eduardo Espinosa Pesqueira
Instituto Nacional de Investigaciones Nucleares, ININ. Mexico
Javier Reyes Trujeque
Universidad Autonoma de Campeche UAC. Mexico
Adrian Velazquez Castro
Museo del Templo Mayor, INAH. Mexico

June 2012

MATERIALS RESEARCH SOCIETY SYMPOSIUM PROCEEDINGS

MATERIALS RESEARCH SOCIETY SYMPOSIUM PROCEEDINGS

Volume 1405E — Advances in Energetic Materials Research, M.R. Manaa, C-S. Yoo, E.J. Reed, M.S. Strano, 2011, ISBN 978-1-60511-382-1

Volume 1406 — Functional Metal-Oxide Nanostructures, A. Vomiero, S. Mathur, Z.L. Wang, E. W-G. Diau, 2011, ISBN 978-1-60511-383-8

Volume 1407 — Carbon Nanotubes, Graphene and Related Nanostructures, Y.K. Yap, D. Futaba, A. Loiseau, M. Zheng, 2011, ISBN 978-1-60511-384-5

Volume 1408 — Functional Nanowires and Nanotubes, K. Nielsch, A.F. i Morral, H. Linke, H. Shin, L. Shi, 2011, ISBN 978-1-60511-385-2

Volume 1409E — Functional Semiconductor Nanocrystals and Metal-Hybrid Structures, K.S Leschkies, P. Nagpal, M.A. Pelton, H. Mattoussi, P. Kambhampati, 2011, ISBN 978-1-60511-386-9

Volume 1410E — Transport Properties in Polymer Nanocomposites II, S. Nazarenko, J. Grunlan, J. Bahr, E. Espuche, 2011, ISBN 978-1-60511-387-6

Volume 1411E — Self Organization and Nanoscale Pattern Formation, S. Persheyev, 2011, ISBN 978-1-60511-388-3

Volume 1412E — Mechanical Nanofabrication, Nanopatterning and Nanoassembly, G. Cross, A. Schirmeisen, A. Knoll, M. Rolandi, 2011, ISBN 978-1-60511-389-0

Volume 1413E — Safety and Toxicity Control of Nanomaterials, W.W. Yu, V.L. Colvin, Q. Dai, P.C. Howard, 2011, ISBN 978-1-60511-390-6

Volume 1415 — MEMS, BioMEMS and Bioelectronics–Materials and Devices, T. Albrecht, M.P. de Boer, F.W. DelRio, M.R. Dokmeci, C. Eberl, J. Fukuda, H. Kaji, C. Keimel, A. Khademhosseini, 2011, ISBN 978-1-60511-392-0

Volume 1416E — Nanofunctional Materials, Nanostructures and Nanodevices for Cancer Applications, S. Svenson, P. Grodzinski, S. Manalis, X J. Liang, W. Lin, 2011, ISBN 978-1-60511-393-7

Volume 1417E — Biomaterials for Tissue Regeneration, C.C. Sorrell, 2011, ISBN 978-1-60511-394-4

Volume 1418 — Gels and Biomedical Materials, F. Horkay, R. Narayan, V. Dave, S. Jin, N. Langrana, J.D. Londono, W. Oppermann, S. Ramakrisha, D. Shi, R.G. Weiss, 2011, ISBN 978-1-60511-395-1

Volume 1419E — Nucleation and Growth of Biological and Biomimetic Materials, P.M. Rodger, J. Harding, L.B. Gower, P. Vekilov, 2011, ISBN 978-1-60511-396-8

Volume 1420E — Multiscale Mechanics of Hierarchical Materials, F. Barthelat, 2011, ISBN 978-1-60511-397-5

Volume 1421E — Three-Dimensional Tomography of Materials, S. Pennycook, 2011, ISBN 978-1-60511-398-2

Volume 1422E — Functional Imaging of Materials–Advances in Multifrequency and Multispectral Scanning Probe Microscopy and Analysis, A. Baddorf, 2011, ISBN 978-1-60511-399-9

Volume 1423E — Dynamics in Confined Systems and Functional Interfaces, M.H. Müser, D.L. Irving, S.B. Sinnott, I. Szlufarska, 2011, ISBN 978-1-60511-400-2

Volume 1424 — Properties and Processes at the Nanoscale–Nanomechanics of Material Behavior, D. Bahr, P. Anderson, N. Moody, R. Spolenak, 2011, ISBN 978-1-60511-401-9

Volume 1425E — Combinatorial and High-Throughput Methods in Materials Science, J.B. Miller, J. Genzer, Y. Matsumoto, R.A. Potyrailo, 2011, ISBN 978-1-60511-402-6

Prior Materials Research Society Symposium Proceedings available by contacting Materials Research Society

Non-destructive Characterization of Cultural Heritage

Mater. Res. Soc. Symp. Proc. Vol. 1374 © 2012 Materials Research Society
DOI: 10.1557/opl.2012.1374

Formation of Hyperspectral Near-Infrared Images from Artworks

Juan Torres[1], Carmen Vega[1], Tomás Antelo[2], José Manuel Menéndez[1], Marian del Egido[2,] Miriam Bueso[2] & Alberto Posse[1]

[1] Grupo de Aplicación de Telecomunicaciones Visuales, Universidad Politécnica de Madrid. Av. Complutense, n° 30, E28040 Madrid, Spain. e-mail: jta@gatv.ssr.upm.es
[2] Instituto del Patrimonio Cultural de España, Ministerio de Cultura, C/ Pintor El Greco, 4, E28040 Madrid, Spain.

ABSTRACT

In this paper, a novel hyperspectral image acquisition system able to obtain a set of narrowband images (~2,25 nm of bandwidth) and the related composition of monochrome images in the near-infrared is described. The aim of this system is to discriminate the materials by their optical spectral response in the range of 900-1700 nm. This system has been developed in the framework of a collaborative project that includes the improvement of the automatic composition of reflectographic mosaics in order to study the underdrawing of large formats big artworks in real-time. The main features of this project are detailed in this paper. Furthermore, a few enlightening results of the hyperspectral system and new lines of research are shown.

INTRODUCTION

The Physics Studies Department of the Instituto del Patrimonio Cultural de España (IPCE) is devoted to the artwork study using non-invasive techniques, such as image processing in several spectral ranges. The conservation and restoration criteria of the cultural heritage demand more and more a minimum intervention in the artwork. For this reason, the Computer Vision systems are very important because, in such systems, invasive sample taking is not required. In addition, they help determining more effectively the sample taking areas.

Among the set of techniques used in this Department, within the VARIM project [1-4] and in collaboration with the Visual Telecommunications Application Group of the Universidad Politécnica de Madrid (UPM), the infrared reflectography has been an increasing issue. Among the reached targets are the construction of an acquisition robot and the complete automation of the reflectographic mosaic composition. Thanks to this development, the study of the underdrawing of big artworks has been possible, in such as a set of Spanish altarpieces from the XV and XVII century without moving or dismantles any painting.

The availability of better sensor devices able to detect wavelengths in the near-infrared (composed by Indium, Gallium and Arsenic - InGaAs) which replace the old ones (composed mainly by lead sulphide) also implies the possibility to develop more precise and faster image acquisition systems.

This fact, and the experience acquired along the last five years, lead to create a new project, composed by the IPCE and the UPM, with the aim of carrying out image spectroscopy within the 900-1700 nm range around 2,25 nm of bandwidth each image.

Thus, a great effort has been carried out developing on the one hand the acquisition system and the composition of the spectral bands and, on the other hand, the acquisitions in the whole work range (900-1700 nm). Traditionally, these last images have been called reflectographic ones in the literature, whereas, the first ones are call hyperspectral images.

A great number of researchers have published papers about the obtained results using image spectroscopy in the last decade, but most of them have explored the visible range of the electromagnetic spectrum. These are mainly multispectral studies, providing a limited number of bands (usually less than 10) and by making use of filters [5-11].

The growing development of the computer capabilities besides the new image acquisition and processing systems provide novel analysis instruments and, additionally, allow to make the old ones more efficient. Thus, the technology becomes more accessible for a great number of restorers, curators, historians and researches in general.

Image spectroscopy applied to the near-infrared may provide two separate results. On the one hand, the different materials which make up an artwork can be distinguished, besides their distribution. This contributes to increase the knowledge about the piece, as well as helping any later conservation and restoration process. On the other hand, the most transparent spectral zone to the infrared radiation can be found out. This contributes to improve the study of the underdrawing in the reflectographic mosaics.

This paper is organized as follows. In the next section, some practical issues about the whole system, the different devices and the software which make it up are explained. Furthermore, this description includes several details about the first trials carried out. Then, some preliminary results we have obtained are described and discussed. Finally, some interesting and brief remarks are gathered in the last section.

EXPERIMENTAL DETAILS

In this section, the physical devices composing the hardware architecture and the software application which handles these devices, focusing on the mosaicing algorithm, are introduced. It is basically composed by two mechanical positioning system (2D and 1D), an InGaAs detector (near-infrared camera), a set of lenses able to diffract the several wavelengths, a lightning system and a PC running a software application in charge of controlling the whole system (as shown in figure 1). Next, the most relevant details about the main devices are explained.

InGaAs detector

The infrared camera is the model XENICS XEVA-FPA-1.7-640-90Hz, which has a InGaAs detector able to work in the near infrared spectral range, between the 900 nm and 1700 nm. It has a thermoelectric cooling system TEI (263 K) which avoids noisy acquisitions. The digital images have 640 x 512 pixels of 20 microns height each one. In addition, the camera is able to acquire images of 14 bits up to 90 fps. The camera is communicated with the PC (or a laptop) via USB. This allows operating it up to 100 m from the artwork.

Hyperspectral set of lenses

The camera mounts a hyperspectral set of lenses (Specim ImSpector Enhanced N17E) . It is a linear device able to diffract the incoming light in 512 lines and project them in one unique gray-scale 2D image in the InGaAs detector. Its spectral range is equal to the detector one and the bandwidth of each line is around 2,25 nm.

4

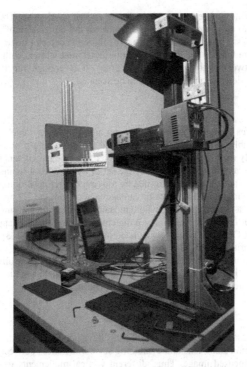

Figure 1. Physical devices with 1D mechanical positioning system

1D Mechanical positioning system

The set of lenses only provides information about one line of the object to be analysed. Thus, a device able to move the camera + the lenses is required. Furthermore, in order to automate as much as possible the process, this device must be controlled by a computer. This system consists of a motorized translation linear guide which allows a vertical linear movement of 40 nm/s. In addition, it can add (that is, load and move) an additional illumination system because it has a charge capacity of 15 kg.

2D Mechanical positioning system

As it has been said, the images acquired by the detector have a resolution of 640 x 512 pixels. Obviously, this is not enough if a high level of detail of the artwork is required. In spite of the fact that this positioning system was designed for the acquisition of reflectographic images, it also allows the acquisition of hyperspectral images when a taking in a not accessible work environment must be done. It is composed by a frame and an easel stand. The frame is a linear 2D guides system able to move up to 40 kg. Furthermore, it has a precision of 40 microns, with a repeatability of 5 microns. The easel stand is able to position the detector up to 2200 mm.

In addition, this structure has two independent multipoint telemetry lasers with a LCD displaying the distance to the artwork with an error of ±1.5mm. It is enough for fixing the required parallelism between the detector and the surface to be scanned.

Varim 2.0

The previous devices are controlled by a computer and a software (developed ad-hoc) called Varim [12]. They are able to make the acquisition and the mosaicing processes in a complete automatic way. The only requirement is to previously set up some parameters (overlapping area, kind of mosaicing, etc.), and the system delivers the set of hyperspectral IR images of some area of an object. Each hyperspectral image is formed by moving the 1D positioning system vertically in order to cover the desired distance. Once this image is acquired, the 2D positioning system is moved on for acquiring a bigger area. The whole mosaic is created by merging consecutive sub-images step-by-step. For any individual union, a matching point in the overlapping area between both images is required. Once this point is calculated in an automatic way [13], the new image is formed using VIPS algorithms [14].

In addition, some useful image processing tools (such as a method for controlling the luminosity in the acquired images [15], a geometrical distortion correction technique [16] and a noise suppression algorithm) are implemented in the software application.

First trials

Once the system was developed, preliminary trials have been carried out. Thus, several tests have been done comparing some lens and evaluating several distances between sample and lens, searching for the best focused image. Thus, different vertical movements were tested until the formation of a whole hyperspectral image without any geometrical distortion was reached. A simple checkerboard of 2x2 cm by square made of graphite over common paper was used for this (see figure 4).

Regarding the spectral response of the system, several trials were made using different substances commonly used to create artworks. Specifically, they are liquid samples such as linseed oil, starch at 20% and white (of an egg), within a test tube.

Other trials were made using oil and acrylic paintings over a rigid surface (tablex covered by white melanin), with a selection of 4 different pigments. Thus, for the oil paintings, the following ones from the Taker Olimpo brand: lamp black (25), yellow light (11), Prussian blue (20) and cadmium red light (44). For the acrylic one: bleu ceruleum (Rembrand 534, series 3), raw sienna (Rembrand 234, series 1), permanent green deep (Lefrank & Bourgeos 0220/350, series 2) and cadmium scarlet (Winsor & Newton, 201, series 4). Regarding the lightning system, the halogen lamp OSRAM (64515-300 w /NAED 58524) was used.

On the other hand, some densitometry measurements were taken using the I-RAD program (developed for a radiography scanner) with the aim of obtaining a graphic showing the distribution of the most and the less absorption areas of the near-infrared spectrum. Nevertheless, a better calibration and understanding of the background heterogeneities is needed.

RESULTS

Once the first trials were carried out, more relevant and promising results were obtained. In this section some of them are showed.

Reflectographic mosaic

Firstly, figure 2 shows a reflectographic mosaic composed by the acquisitions made to a painting of the altarpiece of San Lorenzo de Toro (Zamora – Spain), of 6 meters high. It is composed by 78 images, taken and linked in an automatic way.

Figure 2. a) Photograph of the altarpiece and the mechanical structure for taking the IR mosaic in the capture process (San Lorenzo de Toro, Zamora, Spain). b) Reflectographic mosaic of one the paintings of the altarpiece (*Prendimiento del Papa Sixto*). c) Detail of the mosaic where the names of the colors written by the autor (Francisco Gallego) are clearly visible.

Hyperspectral images

With every material used as background for the different trials, a discontinuity has been observed in the spectral density distribution. As shown in figure 3, this fact is reflected as a darker thin band in close to the centre of the hyperspectral image. Although it has not been completely checked, we assume that there may be some slight diffraction problem with the lenses. Apart from that, the limits of our system, as can also be seen in figure 3 are, in practice and according to the calibration made, between 990 and 1730 nm.

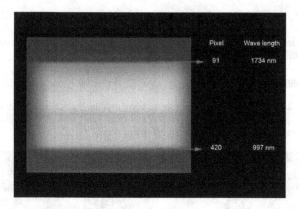

Figure 3. Background hyperspectral image.

Bands reconstruction (gray level images)

Figure 4 also shows the result of the acquisition process. We can obtain two different image informations: a) hyperspectral images in the range of 990 to 1730 nm; b) monochrome images, composed by linking information of the same wavelength of many narrow bands acquired in different camera sequential positions.

As it can be seen, the obtained monochrome images have no geometrical aberration. For obtaining that, the hyperspectral images were acquired by moving the camera in a vertical way in 100 microns intervals. The number of acquired images was 300, and the field of view was 6 cm x 3 cm. In the same figure the appearance of the hyperspectral images in two different acquired lines is highlighted.

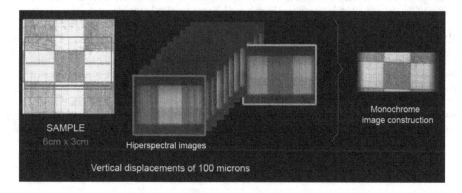

Figure 4. Example of both hyperspectral images and reconstructed monochrome images.

Oil, egg albumen and starch

Figure 5 shows one of the acquired hyperspectral images of the liquids in three different test tubes. The infra-red absorption bands that the linseed oil produces can also be clearly seen. Two bands were selected (1153 nm and 1471 nm) for the reconstruction of monochrome images, where the different gray level output can be noticeable. Figures 6 and 7 represent the numerical data obtained for these three samples.

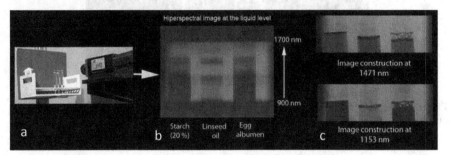

Figure 5. Hyperspectral image of the experiment made with linseed oil, egg albumen and starch. a) Photograph of the experimental setup. b) Hyperspectral image at the liquid level. c) Monochrome images at 1153 and 1471 nm.

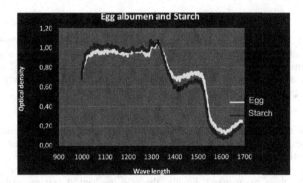

Figure 6. Optical density profiles of the hyperspectral samples obtained from the egg albumen and starch, in the range of 990 to 1730 nm.

9

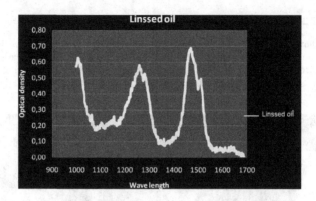

Figure 7. Optical density profile of the hyperspectral sample obtained from the linseed oil, in the range of 990 to 1730 nm.

<u>**Acrylic painting**</u>

Figure 8 shows one of the acquired hyperspectral images with some acrylic samples, and two monochrome images of two selected bands where the blue and green samples (1186 nm and 1660 nm) show greater spectral difference. Figure 9 shows the optical density profiles of these results, highlighting their difference.

<u>**Oil painting**</u>

The same experiment was done using oil painting samples instead of acrylic ones. Figure 10 shows the results of the oil painting samples. The hyperspectral images have been acquired in the middle of the strips, and were digitally corrected in order to enhance the contrast. Thus, it can be clearly seen some absorption bands for the yellow pigment. Figures 11 and 12 show these results.

DISCUSSION

Related to the reflectographic mosaics, the camera with the InGaAs detector provides lower-noise images (compared to the old Vidicon detectors) so, the quality is increased. This makes easier the automation of the mosaic composition and minimizes the image processing stages. Nevertheless, the resolution of the new images is less than the Vidicon one; hence a greater number of images must be acquired. Finally, the new mechanical systems make faster the acquisition task.

Concerning the hyperspectral images, at this moment, the number of vertical movements for obtaining the hyperspectral images is around 300, but increasing this number the covered area can be higher.

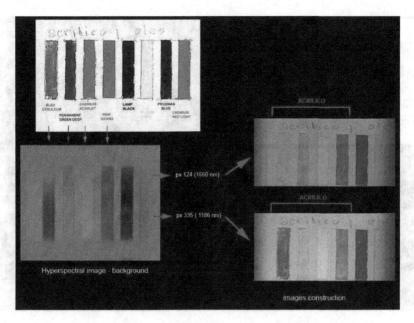

Figure 8. Acrylic samples used, hyperspectral image and 2 monochrome images (at 1186 and 660 nm) obtained.

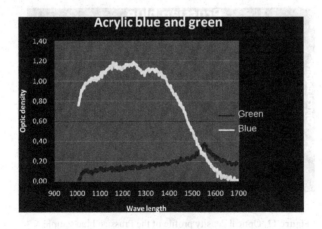

Figure 9. Optical density profile for the acrylic blue and green samples.

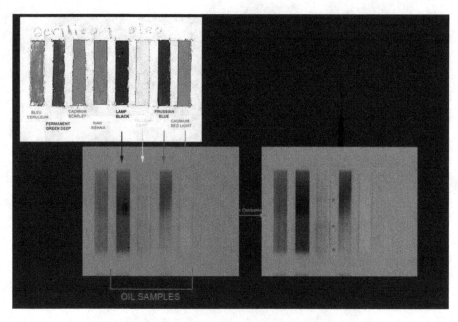

Figure 10. Hyperspectral image of the oil painting samples.

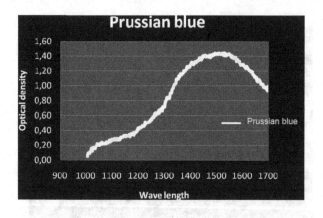

Figure 11. Optical density profile of the Prussian blue sample.

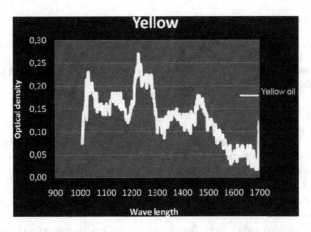

Figure 12. Optical density profile fo the light yellow oil painting sample.

The observed results in the oil sample show that it is possible to see absorption bands produced by this substance and its response is completely different to the white and starch ones. In these cases, the main component is water. The volume of substance exposed to the infrarred radiation makes easier the observation, and the test tube material does not seem to interfere in the result. Furthermore, the monochrome images clearly show the different spectral response profiles of these substances.

Other interesting result is the spectral profiles of the painting samples used in the trials, which, as shown, are very different. Not only in specific ranges of the spectrum, but also in absorption narrow bands. Furthermore, in spite of the infrared radiation, as it is known, is able to go through the painting layers making these transparent to that radiation.

On the other hand, the acquired images have 8 bpp. Our current work goes in the line of acquiring and managing images of 14 bpp. These will improve the precision of the results, working with 16384 gray-levels (much more than the current 256).

In addition, other issue that has to be addressed in the near future is a rigorous study of hyperspectral images of painting patterns made ad-hoc, using old and modern materials, and also other materials different from the ones used in paintings.

Finally, the inclusion in the study of visible and ultraviolet ranges is also considered, just to complete the spectrum information about the materials and the evolution of the profiles.

CONCLUSIONS

At this moment, the trials carried out stimulate our future research. We hope to provide a useful tool for the artwork study which helps to the understanding of goods and the restoration and conservation tasks within the research lines of non-invasive techniques.

ACKNOWLEDGMENTS

Authors wish to explicitly thank Araceli Gabaldón for being part of the soul of this complete project, from the IR automation system to the hyperspectral acquisition one. They also thank to Enrique García and the companies, Ana Manzanares, Greenlight Solution S. L. and Infaimon S. L. for their collaboration.

REFERENCES

1. T. Antelo, M. Del Egido, A. Gabaldón, C. Vega and J. Torres. *El proyecto VARIM: Visión Artificial aplicada a la Reflectografía de Infrarrojos Mecanizada* in *Innovación tecnológica en conservación y restauración del Patrimonio. Tecnología y Conservación del Patrimonio Arqueológico I*, Universidad Autónoma de Madrid, Madrid, 2006, 87-99.

2. J. Torres; A. Posse, J.M. Menéndez, A. Gabaldón, C. Vega, T. Antelo, M. Del Egido and M. Bueso, *e_conservation* 4 (2008) 27-42.

3. T. Antelo, A. Gabaldón and C. Vega, *Bienes Culturales. Ciencias aplicadas al Patrimonio 8*, Ministerio de Cultura, Madrid, 2008, 61-73.

4. J. Torres A. Posse and J.M. Menéndez, *Bienes Culturales. Ciencias aplicadas al Patrimonio 8*, Ministerio de Cultura, Madrid, 2008, 89-97.

5. S. Baronti, A. Casini, F. Lotti, S. Porcinai, *Applied Optics* 37, 8, (1998) 1299.

6. C. Fischer, I. Kakoulli, *Reviews in Conservation*, 7 (2006) 3-16.

7. M. Bacci, *Archeometriai Mühely*, 2006/1, 46-50.

8. M. Kubik, Hyperspectral Imaging: A New Technique for the Non-Invasive Study of Artworks in *Physical Techniques in the Study of Art, Archaeology and Cultural Heritage*. Vol. 2, Creagh and Bradley eds., Elsevier. Amsterdam, 2007, 199.

9. A. Pelagotti, A. Del Mastio, A. De Rosa , A. Piva, *Signal Processing Magazine IEEE* 25 (2008) 27-36.

10. A. Ribés, R.Pillay, F. Schmitt and C. Lahanier, *Signal Processing Magazine IEEE* 25 (2008) 14-26.

11. S. Giró, Sistemas Digitales de Adquisición de Imágenes Visibles, Infrarrojos e Hiperespectrales, in *La Ciencia y el Arte II*. M.A. del Egido & D. Juanes (eds,), Secretaría General Técnica. Ministerio de Cultura, Madrid, 2010, 205.

12. VARIM Project: http://www.gatv.ssr.upm.es/wikivarim

13. A. Posse, J. Torres, J.M. Menéndez, *Matching points in poor edge information images*, 16th IEEE International Conference on Image Processing, El Cairo, Egypt, 2009, 197-200.

14. K. Martinez, J. Cupitt, *VIPS – a highly tuned image processing software architecture*, IEEE International Conference on Image Processing, Genoa, Italy, vol 2, 2005, 574-577.

15. J. Torres, J.M. Menéndez, *An adaptive real-time method for controlling the luminosity in digital video acquisition*, AESTED International Conference on Visualization, Imaging and Image Processing, 2005, Benidorm, 133-137.

16. J. Torres, J.M. Menéndez, *A practical algorithm to correct geometrical distortion of image acquisition cameras*, IEEE International Conference on Image Processing, vol III, Singapur, 2004, 2451-2454.

... ...
... ...

Mater. Res. Soc. Symp. Proc. Vol. 1374 © 2012 Materials Research Society
DOI: 10.1557/opl.2012.1375

Study of Colonial Manuscripts from San Nicolás Coatepec, México, Through UV&IR Imaging and XRF

Jocelyn Alcántara García[1], José Luis Ruvalcaba Sil[2] and Marie Van der Meeren[1]

[1] Coordinación Nacional del Patrimonio Cultural, Instituto Nacional de Antropologia e Historia (INAH). Ex Convento de Churubusco s/n esq. General Anaya. Col. San Diego Churubusco, México D. F. 04120, México. e-mail: jocelynag@gmail.com

[2] Instituto de Física, Universidad Nacional Autónoma de México (UNAM), Apdo. Postal 20-364, México D. F. 01000, México- e-mail: sil@fisica.unam.mx

ABSTRACT

The necessity of studying cultural heritage through non-invasive and non-destructive techniques has led to significant advances in the last decade. One of the most recent advancements in this theme in Mexico is the portable X-ray system SANDRA, which was used to study three manuscripts directly related to the history of "San Nicolás Coatepec", Mexico. X-ray fluorescence was chosen as the suitable technique because it can provide a fast qualitative and quantitative multielemental high sensitivity analysis. The documents were examined globally, using imaging techniques with UV and IR lighting. This research evinced a change in the composition and evolution of writing materials (inks and pigments) and provided information concerning historical use of the documents and its actual legal value as a property document. It also stressed the need of spanning these results to an extensive research attaining other regions of Mexico, in order to fully understand the Mexican documents particularities, aging and deterioration. This, in turn, will provide not only historical material information but also an invaluable scoop to understand deterioration and conservation issues.

INTRODUCTION

Although stylistic and aesthetic considerations can provide answers concerning provenance and age of objects, these cannot be taken as conclusive [1]. Some specialists can emulate styles and physical appearances. However, aging processes, constituent materials and their characteristic compositions are unique of every singular historical or artistic piece. In this sense, analyses are fundamental to determine singularities of the objects. Of special importance are the non-destructive techniques, not only because they can provide useful and accurate results but also because they further allow later studies without altering the original material.

Among the usual techniques for analysis of works of art and historical items, the spectral imaging and XRF are of common usage [2-10]. Accordingly, we used UV and IR-imaging combined with XRF to study three colonial manuscripts from the Mexican town "San Nicolás Coatepec". Two of the documents are dated in the 16th century, whereas the third is thought to be from the 19th century.

Written in Spanish, the manuscript named "Mercedes" is a legal one, where it is stated that the Viceroy of New Spain gave these lands to the people of the town.

On the other hand, The San Nicolás Coatepec Codex (códice de Coatepec) is an original manuscript that relates in Nahuatl the foundation of "Coatepec de las Bateas". In this document, it is stated both the efforts of Nicolás Miguel to build the San Nicolás Temple and how the limits of the town were established. The third document is a colorfully translated version of the latter (to Spanish).

In 2011 this set of manuscripts had to be kept for restoration processes at the National Coordination of Conservation and Restoration (CNCPC) of the Instituto Nacional de Antropologia e Historia (INAH). As part of the process, we examined them to obtain insights both on their constituent materials and –possible– degradation processes.

EXPERIMENTAL

Prior to restoration processes, the documents were studied at the Institute of Physics (IF-UNAM) using XRF and imaging. A detailed record of the manuscripts was performed using infrared imaging (IR) with an IR filter (87C polyester). Ultraviolet images (UV) were taken without filter using 365 and 254 nm wavelength hand lamps. The documents were also studied using visible light.

After the global examination by the imaging techniques, the inks and paper of the document were analyzed using the portable XRF SANDRA device [2] with a 75W Mo X-ray tube and a Si-PIN X123 detector from Amptek. XRF conditions for the X-ray tube were 40 kV and 0.6 mA, 120 seconds per region. A total of 231, 135 and 156 measurements were carried out on Mercedes, San Nicolás Codex and the 19th century translation, respectively. Through the measurements, it was possible to analyze inks, paper and pigments (when existent). The beam spot was 1 mm diameter.

DISCUSSION

Coatepec Codex (Códice Techialoyan de San Nicolás Coatepec)

At first, this document (Figure 1) was thought to be the oldest of the three studied. However, recent historical findings revealed that during the 18th century in Mexico, it was common practice to write apocryphal documents, *e.g.* the date of recently generated 18th century documents was different, even centuries earlier. The good state of preservation of this Nahuatl-written document is remarkable. Besides some wrinkles, broken corners, partial loss of book binding and hardening of the leather, the document can be easily manipulated without risks.

The same deterioration (stains, mostly present in the inner part of folios) was visible both with visible and UV light. No previous restoration processes were detected.

The UV and IR imaging suggest presence of carbon inks through the entire manuscript: opaque both under UV illumination and IR. However, the XRF results revealed relatively high contents of sulfur and iron, which is expected for iron-gall ink presence [11].

Besides sulfur and iron, the document's inks posses magnesium, chromium, chloride, potassium and calcium as main elements, and as traces (or in few measurements) lead, copper and zinc. According to previous studies [1, 8, 10-14], iron-gall inks are mainly composed of $FeSO_4$ and organic material. Minor constituents are principally (but not always found in European studies) Mn, Cu and Zn [3]. As for the paper, Ca, K and partially S can be associated to fillers and sizing materials (starch, $Ca.SO_4$, glue, etc.) [10].

On the other hand, it is known that the thickness of ink layers on a manuscript varies considerably from one measurement to the other. Therefore, the effect of X-ray intensity increment due to higher volume of ink in the paper has to be taken into account [3, 8]. To evaluate migration of metals from inks to paper, the comparison of the analyzed regions is through metal to iron ratio (e.g. Cu/Fe, Mn/Fe, etc.), whereas to identify possible degradation mechanisms S/Fe and S/Ca are discussed [10].

Figure 1. Two pages of the Coatepec Codex or Códice Techialoyan de San Nicolás Coatepec.

Mn/Fe maximum contents are the same for paper and inks (0.36) while Cu/Fe ratio in paper is ca. half of the maximum content in inks (0.36 and 0.17, respectively). No significant ratios of other block d metals were detected. Similarities in all the estimated ratios for ink and paper measurements can be related to a high migration process, which in turn is commonly associated to highly degraded paper supports. The preservation state of the Codex is good, though.

The S/Ca and S/Fe peak area ratios calculated for both inked and uninked areas of the three documents are in Table 1. When the inked areas posses a S/Fe ratio of <1, the excess of iron is associated as a major contributing factor for the corrosion process. When S/Fe>1, acidity is thought to be the mayor contributing factor of deterioration. [10, 15] On the other hand, high S/Fe ratio in paper is a result of low concentration of Fe. Therefore, the S presumably comes from the calcium sulfate ($CaSO_4$) used as filler in the paper.

Table 1. Average, maximum and minimum S/Ca, Fe/Ca and S/Fe ratios for studied manuscripts.

Inks

	S/Ca			Fe/Ca			S/Fe		
	avr	max	min	avr	max	min	avr	max	min
Codex	0.19	0.38	0.09	0.14	0.52	0.07	1.54	2.75	0.47
Mercedes	0.32	0.07	3.03	1.11	0.20	2.96	0.35	0.06	3.63
Translation	0.53	1.70	0.00	1.05	3.80	0.12	0.87	14.37	0.00

Papers

	S/Ca			Fe/Ca			S/Fe		
	avr	max	min	avr	max	min	avr	max	min
Codex	0.17	0.33	0.08	0.14	0.32	0.07	1.33	2.49	0.61
Mercedes	0.28	0.46	0.09	0.65	1.44	0.22	0.51	0.96	0.09
Translation	0.23	3.95	0.00	0.53	1.05	0.29	1.24	2.55	0.34

As shown in Table 1 by the Fe/S ratio, the document has areas that are from very acidic to endangered by Fe-excess. However, the high contents of Ca both for paper and ink, along with almost neutral pH measurements, justify the stability of cellulose towards corrosion.

When comparing the iron and the sulfur relative contents to previously reported iron-gall inks, it is evident that for the San Nicolás Codex both are considerably smaller. This evidence with the UV imaging relates to a mixture of carbon with iron gall ink. Therefore, the similarities of what seemed to be an advanced migration process from metals of inks to paper, is actually proof of almost identical contents of the elements present in the inks and the paper.

As mentioned earlier, the main inorganic components in iron gall inks and in paper are Fe and Ca, respectively. In Figure 2 it is shown the distribution of the sulfur and iron contents in the 16th century dated manuscripts. The distribution of measurements evinces similarities in the composition of most of the inks of the Codex. This, in turn, implies the usage of the same recipe of preparation, small manufacturing time and/or use of only one ink through the entire process. If we think of this document as written –perhaps– by only one person (an "official writer") during a relatively short period of time, it seems logic to find almost identical inks through the entire document.

Mercedes (Merced de tierras otorgado por el Virrey de la Nueva España)

Written in Spanish, the state of preservation of this 1563 document is precarious both for paper and the leather bookbinding. The damages present span from small fractures of certain zones to severe loss of fragments (many of which could not be located). However, except for offsetting to neighboring pages, no evidence of ink-corrosion was found.

Under UV and IR light the inks show typical behavior for iron gall inks [11]: very opaque under UV illumination and slightly gray in IR imaging (Figure 3). At first sight, most of the inks have similar opacities except for two folios, which have very different calligraphy. No previous restoration processes or carbon inks were detected. Nevertheless, UV imaging allowed unequivocal identification of severe damages due to high humidity, as are tidelines and both fungal and water stains.

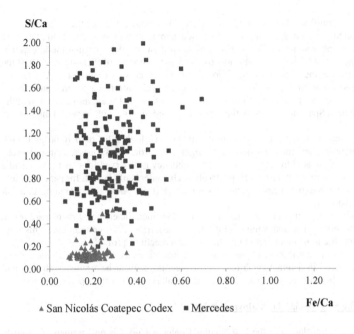

Figure 2. S/Ca and Fe/Ca ratios in the 16th century documents.

Figure 3. Folio 11 (verso) of Mercedes under visible (left), IR (center) and UV (right) light.

XRF allowed identifying high contents of Fe, S, Cl, K and Ca in all the measurements, which was expected for iron gall inks in a bleached paper. From these, approximately 2 % contain also significant amounts of Zn and Cu as main components, which when found together are in approximately in a 4:1 ratio. We also identified high proportions of Pb and Hg in 10 of the 228 measurements (in the signatures of folio 15 both elements were present). Unlike the previously discussed manuscript, no Mn was identified, which, according to previous studies make these inks "more European" than those of San Nicolás Codex. Taking into account that the document is written in Spanish, it is likely that Spaniards used to write with imported inks during the 16th century.

As seen in Table 1, inked areas span from low to very high S/Fe ratio (0.06 and 3.63, respectively), which implies endangerment that ranges from iron excess, passes through well-balanced inks and ends in acid hydrolysis. This same tendency is observed for paper (from acidic to areas of high content of iron). However, relatively high contents of Ca both for paper and ink, along with almost neutral pH measurements (average pH is 6), allow understanding that ink corrosion is not so advanced.

On the other hand, significant contents of the most abundant metallic elements in inks are found in paper, which involves important migration processes occurring. As a consequence, this high mobility of the metallic ions might yield to pseudo-Fenton reactions [16].

Figure 2 shows the distribution of the sulfur and iron contents relative to calcium. Since distribution is clustered in a relatively broad region, seems likely to be written with few and similar ink recipes, although perhaps with different raw materials over a short period of time.

Translation of the San Nicolás Techialoyan Codex

The 19th century translation of the San Nicolás Coatepec Codex is also written in Spanish. Since it is colorfully illustrated, the present discussion will be divided in black inks and in the pigments with which it was illustrated. In general, the preservation state is remarkable through the entire manuscript, and except for some masking tape and glued paper in a map located in the central pages (see Figure 4), no previous restoration processes were identified neither with visible or any light.

All inks present are opaque under UV light and slightly grey in IR imaging, which is typical for iron-gall inks [11]. Furthermore, several opacities of grey were observed through the later, pointing to utilization of different inks.

The elements existent as main components are S, K, Ca, Fe, Mn, Cu and Zn. As minor components Cl and Ti were detected, and in only few of the measurements Ba, Pb, Cr or Hg were identified. A comparison of the analyzed inks using the most abundant elements is shown in the ternary diagram of Figure 5. Three main cluster regions are identified, being the mayor one that with an ink composition of Cu/Fe, Zn/Fe and Mn/Fe of approximately 0.6, 0.4 and 0.3, respectively. These regions are followed by inks with no Cu and finally by inks with no Zn in its composition.

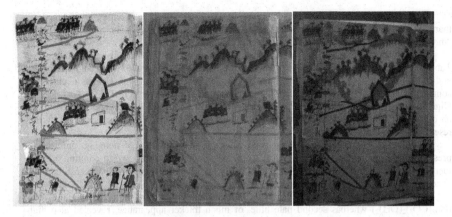

Figure 4. Folio 15 (map) from translation under visible (left), IR (center) and UV (right) light.

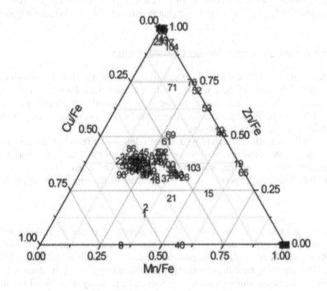

Figure 5. Ternary diagram of X-ray ratios of the main metallic elements in inks from 19th century translation.

The very complex mixture of inks could imply usage of different ink recipes, involvement of many people in the manufacture and/or that it took a long time to finish the translation (many "moments" through the history of the document).

In this manuscript the Fe/S ratios for inks encompass very acidic and well-balanced inks (see Table 1). Concerning the paper, there are both zones with a marked proclivity to acid hydrolysis and oxidation due to Fe. Migration of most of the elements present in the inks has occurred at some level, further endangering the cellulose to pseudo-Fenton reactions [13,16].

High contents of Mn found in this translation and the corresponding Nahuatl codex seems to be an outstanding characteristic of Mexican inks. This hypothesis is currently under study by our research group [17].

Attaining the pigments, no colored zone was transparent under UV imaging, pointing to presence of inorganic pigments (see Figure 4) [5, 18]. Nevertheless, it is worth pointing out that identification of pigments was performed based only on relative contents in XRF measurements. A complementary technique is desirable for results ascertainment.

Three tones of blue are existent: a lighter blue, of high Cu-contents is thought to be azurite ($Cu_3(CO_3)_2(OH)_2$) whereas second pale blue, of much thicker appearance, revealed also high contents of S, Zn and Ba. Therefore, the presence of a Lithopone base of preparation seemed probable (ZnS with $BaSO_4$). Darker blue was identified as Prussian ($Fe_4[Fe(CN)_6]_3$).

Yellowish green was identified as chrome green, a mixture of Prussian blue with lead chromate ($PbCrO_4$) due to significant contents of Fe, Pb and Cr. The red-orange is thought to be red lead (Pb_3O_4) and finally, it seems reasonable that purple is Lithopone with an organic purple colorant.

Evidence of 19th c. usage in the San Nicolás Coatepec Codex

In several pages of the San Nicolás Codex there were found distinctive marks with the shape of a cross or numbers. The color of these was very different from the color of the used inks in the supposed 16th century Codex, and similar to the 19th c. translation (black and brownish, respectively). Furthermore, imaging revealed these marks as UV opaque and IR gray (of different opacities), as in the 19th c. document.

Comparative analysis of the inks employed in both documents revealed in many cases that the ink-mark in the earlier document was the same with which the Translation was produced. Since a correspondence of the original and translated folios exists when it is the same ink, it is feasible that the "translator" or the person studying the earlier manuscript put a mark in the later.

CONCLUSIONS

To the best of our knowledge, the set of three manuscripts studied here are unique in terms of the close relationship to a place and its history. It was also a unique opportunity to study the evolution of inks in one Mexican place, the possible degradation mechanisms and stability. The results revealed what seems to be a fingerprint for iron gall inks produced in Mexico: high Mn-content. Moreover, the present study evinced the importance of studying in detail the inks with which the documents to be restored were composed. The earlier states for the need of finding specific solutions to specific conservation issues, which quite often differ dramatically from previously found problems (*i.e.* in Europe). Our research group has currently a study of this nature on course.

ACKNOWLEDGMENTS

Authors acknowledge to the people and authorities of San Nicolas Coatepec, Estado de Mexico, Mexico, for their interest to perform this study. This research has been carried out with support of UNAM PAPIIT IN403210 project as well as the CONACYT Mexico grant 131944. This study has been performed in the frame of the Non Destructive Study of the Mexican Cultural Heritage ANDREAH network (www.fisica.unam.mx/andreah) in collaboration of the paper workshop of the Coordinación Nacional del Patrimonio Cultural, Instituto Nacional de Antropologia e Historia (INAH).

REFERENCES

1. M. Mantler, M. Schreiner, *X-Ray Spectrometry* 29 (2000) 3.
2. J. L. Ruvalcaba Sil, D. Ramírez Miranda, V. Aguilar Melo, F. Picazo, *X-Ray Spectrometry* 39 (2010) 338.
3. M. Malzer, O. Hahn, B. Kanngießer, *X-Ray Spectrometry* 33 (2004) 229.
4. D. Ambrosini, C. Daffara, R. Di Biase, D. Paoletti, L. Pezzati, R. Bellucci, F. Bettini, *Journal of Cultural Heritage* 11 (2010) 196.
5. J. K. Delaney, E. Walmsley, B. H. Berrie, C. F. Fletcher in *Scientific Examination of Art: Modern Techniques in Conservation and Analysis,* Proceedings of the National Academy of Sciences, Washington D.C., 2010, 120-136.
6. E. Carretti, M. Milano, L. Dei, P. Baglioni *Journal of Cultural Heritage* 10 (2009) 501.
7. A. Duran, M. L. Franquelo, M. A. Centeno, T. Espejo, J.L. Perez-Rodriguez, *J. Raman Spectroscopy* 42 (2011) 48.
8. O. Hahn, W. Malzer, B. Kanngießer and B. Beckhoff, *X-Ray Spectrometry* 33 (2004) 234.
9. D. Bersani, P.P. Lottici, S. Virgenti, A. Sodod, G. Malvestuto, A. Botti, E. Salvioli-Mariani, M. Tribaudino, F. Ospitali, M. Catarsi *J. Raman Spectroscopy* 41 (2010) 1266.
10. M. Trojan-Bedynski, F. Kalbfleisch, S. Tse, J. Sirois, *JACCR* 28 (2003)1.
11. B. Reißland, J. Hofenk de Graff, *Condition rating for paper objects with iron-gall ink,* Netherlands Institute for Cultural Heritage Report, Amsterdam, 2000.
12. B. Kanngießer, O. Hahn, M. Wilke, B. Nekat, W. Malzer, A. Erko, *Spectrochimica Acta B* 59 (2004) 1511.
13. J. Kolar, M. Strlic, M. Budnar, J. Malesic, V. Simon Selih, J. Simcic, *Acta Chim. Slov.* 50 (2003) 763.
14. V. Rouchon, C. Burgaud, T. P. Nguyen, M. Eveno, L. Pichon, J. Salomon *Papier Restaurierung* 9 (2008) 18.
15. M. C. Sistach, I. Espadaler *Organic and Inorganic Components of Iron Gall Inks* in preprints, 10th Triennial Meeting, ICOM Comitee for Conservation 2 (1993) 485.
16. J. Kolar, A. Stolfa, M. Strlic, M. Pompe, B. Pihlar, M. Budnar, J. Simcic, B. Reißland *Analytica Chimica Acta* 555 (2006) 167.
17. J. Alcantara García, J. L. Ruvalcaba-Sil, M Van der Meeren, 2011 (unpublished results).
18. M. L. Gómez, *Examen científico aplicado a la conservación de obras de arte*, Cátedra Cuadernos de Arte, Madrid, 2004, 51-63.

Technical Studies in Art History

Mater. Res. Soc. Symp. Proc. Vol. 1374 © 2012 Materials Research Society
DOI: 10.1557/opl.2012.1376

Chromatographic Investigations of Purple Archaeological Bio-Material Pigments Used as Biblical Dyes

Zvi C. Koren
The Edelstein Center for the Analysis of Ancient Artifacts, Department of Chemical Engineering, Shenkar College of Engineering and Design, 12 Anna Frank St., 52526 Ramat-Gan, Israel.
e-mail: zvi@shenkar.ac.il

ABSTRACT

This article discusses recent scientific research performed by the author in understanding the composition of archaeological purple pigments and dyes from molluskan sources, which were primarily used for the dyeing of royal and priestly textiles, as also cited in the Bible. Towards this end, the high-performance liquid chromatography (HPLC) method has been applied to the qualitative and quantitative multi-component fingerprinting of purple pigments extracted from various *Muricidae* mollusks inhabiting the Mediterranean waters. The results show that the colorants in these purple pigments belong to three chemical groups: the indigoids (of major importance), the indirubinoids, and the isatinoids. Application of this analytical method to purple pigments and dyes on archaeological artifacts from the ancient Near and Middle East has lead to a number of breakthroughs and discoveries made by this laboratory. These include the following: decipherment of the optimal method by which the ancients practiced purple-dyeing by completely natural means; first HPLC analysis of a raw unprocessed purple archaeological snail pigment and the resulting identification of a dibrominated indirubin in this pigment; discovery of the purple pigment as the sole paint pigment on a 2,500 royal marble jar from the Persian King Darius I; and the discovery that a 2,000 year old miniscule fabric found atop the Judean Desert palatial fortress of Masada belonged to the royal purple mantle of King Herod I and is the first Biblical Argaman dye found in ancient Israel.

INTRODUCTION

The analysis of cultural heritage textile dyes and pigments from organic sources – flora and fauna – requires the use of micro-sampling in order to extract the maximum information regarding these colorants. Non-destructive investigations have been shown to be excellent for the identification of inorganic pigments. However, they are at best limited in their ability to produce a full scientific fingerprinting of natural organic dyestuffs, which contain numerous components. The optimal technique for studying organic dyes and pigments requires a separation method, especially HPLC (high-performance liquid chromatography) followed by spectrometric detection of the visible- and UV-absorbing components. The latter is best performed with a photodiode array (PDA) detector, which can then be linked to a mass spectrometer (MS) for the identification of hitherto unknown colorants.

This paper will address the recent discoveries of purple pigments produced from various molluskan sources found in textiles and objects from more than two millennia ago. The purple of the ancients is undoubtedly the most fascinating and mystifyingly complex pigment of all the natural colorants investigated. In the past two decades, this molluskan pigment has been the focus of increased research. Classical authors, such as the 4th century BCE Greek philosopher Aristotle [1] and the 1st century CE Roman historian Pliny [2] have written on it, and more recently Cardon [3, 4] and Haubrichs [5, 6] have reviewed its history. The chemistry of this purple pigment has been reviewed [7, 8], and analytical methods have been developed for multi-component identifications of *Muricidae* pigments via liquid chromatography [9 - 16].

These purple pigments were extracted from the hypobranchial glands of certain *Muricidae* sea snails inhabiting the Mediterranean and nearby waters. The three main molluskan species that have been associated with purple dyeings in the Mediterranean region are [3 - 5, 17, 18]: (a) *Hexaplex (= Murex = Phyllonotus = Trunculariopsis) trunculus*; (b) *Bolinus (= Murex = Phyllonotus) brandaris*; and (c) *Stramonita (= Purpura = Thais) haemastoma*. They are shown in Figure 1. The extracted colorants were used as a paint pigment and primarily as a textile dye, though the former usage probably chronologically preceded the latter. Purple and violet garments bestowed upon the owner an aura of power and sacredness and, thus, these textiles were the prerogative of sovereigns, military generals, eminent officials, and high priests.

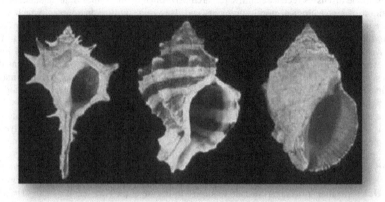

Figure 1. Three *Muricidae* sea snails inhabiting the Mediterranean Sea and processed in antiquity to extract the purple pigment (from left to right): *Bolinus brandaris*, *Hexaplex trunculus*, and *Stramonita haemastoma* (courtesy of the Eretz Israel Museum, Tel-Aviv).

It is interesting to note the interconnection between the Hebrew and Greek versions of the two Biblical forms of purple. As such, the names of these two "purples" mentioned in the Hebrew Bible are first cited in Exodus 25:4 and are denoted in Hebrew as *Tekhelet* and *Argaman*. According to traditional accounts, the first translation of the Hebrew Bible into another language – Greek – was begun in the 3rd cent. BCE. This literary project was undertaken, according to legend, by about seventy Greek-versed Rabbis, and thus this body of work is known as the Septuagint (or Septuaginta). In that translation, *Tekhelet* is rendered as υακινθον (yakinthon), that is, a hyacinth color, and Argaman as πορφυραν (porfyran), a purple. There is an interesting Talmudic legend, dated from sometime in the first two centuries CE, regarding this legendary venture, and it is described in the Babylonian Talmud (Tractate Megillah 9a), as follows:

"*It is related of King Ptolemy that he gathered seventy-two Elders and placed them in seventy-two [separate] houses, without revealing to them why he gathered them. He entered each one's house and said to them: Write* [i.e., translate] *for me the Torah of Moses your master. God then gave wisdom in the heart of each one and they all concurred on one identical erudition* [i.e., translation]".

The references to these two Biblical chromatic names are with respect to the colors of woolen dyeings produced from the pigments extracted from certain sea snails. The exact colors of these two textile dyes has been in dispute for nearly two millennia, though the past century has seen more intense discussions on this topic, thanks to the advances made by modern analytical instrumentation.

This laboratory has conducted intensive research on the purple molluskan pigment for nearly the last two decades in order to decipher the colors and chemical constitutions of these two Biblical dyes as well as to have a better scientific understanding of the production and usage of this pigment. The research strategy involves a multistage approach consisting of the following steps:

- Extraction of the purple pigment from the snail;
- Dissolution of the sample with an optimal solvent;
- Development of an analytical HPLC separation method for the colorants constituting the pigment;
- Application of the method to modern and archaeological pigments;
- Determining the malacological provenance of the ancient pigment.

MURICIDAE FAMILY OF SEA SNAILS: EXTRACTION OF THE PURPLE PIGMENT

It has been known for some time now that the *H. trunculus* produces considerably more pigment than either *B. brandaris* or *S. haemastoma*. In addition, whereby the latter two produce reddish purple pigments that can be commonly referred to as crimson, maroon, Bordeaux, etc., the *H. trunculus* can produce a bluer purple or violet pigment, but this snail (or a very closely related species) can also produce red-colored pigments, similar in color to the other two sea snails.

The chromogenic precursors to the final purple pigment are actually colorless in the hypobranchial glandular fluid of the live snail. This gland is reachable by breaking the snail's shell with a hard object (e.g., stone or hammer) and exposing the vein, as shown in Figure 2.

Upon excising the gland with a sharp object (knife or scissors) or simply puncturing it, the enzyme that is present in a different compartment of the gland comes in contact with the precursors. Consequently, in the presence of air and light, this enzyme transforms the colorless precursors through various complex photo-oxidative processes to the final purple pigment [7]. The first short-lived color of the exposed chromogens is white, which then turns to yellow, then green, and finally purple, and some of these spontaneous color stages are depicted in Figure 3. It should be noted that the development of the final purple color stage is greatly expedited if exposed to direct sunlight as opposed to just room lighting conditions.

The next step in the research stage is to qualitatively and quantitatively analyze this pigment.

Figure 2. Exposed hypobranchial gland, seen as a gray vein, in a *H. trunculus* snail (© Zvi C. Koren).

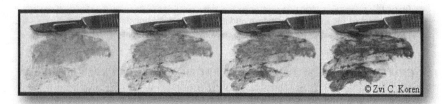

Figure 3. Color development of the precursors exposed to air and light, from left to right; the first three images are of the color development in room light in about 10 minutes and the last image of the sample exposed to the sun for a few minutes (© Zvi C. Koren).

SOLVENT SEARCH: FINDING THE OPTIMAL SOLVENT FOR THE DISSOLUTION OF PURPLE PIGMENTS

In order to perform any chromatographic analysis via the HPLC method, the sample must be dissolved prior to its injection into the instrument. The major problem with the water-insoluble purple pigment, which has been shown to consist of various indigoids and related colorants [19], is that these compounds are only sparingly soluble even in hot organic solvents. Previously, in order to dissolve the primarily indigoid-based pigment, glacial acetic acid, pyridine, and dimethyl formamide (DMF) were used for this purpose. However, this laboratory has found that the use of dimethyl sulfoxide (DMSO) at the elevated temperature of 150 °C is the optimal solvent for these pigments. It was already observed that it is an excellent solvent for red safflower dyeings [20, 21], which lead to its application also to indigoids [22]. This solvent has a high normal boiling point (189 °C) and has very little health and safety issues. It was found that this high temperature for the duration of 5 minutes increases the dissolution of the sparingly soluble dibromoindigo and, when performed under subdued lighting conditions, does not cause any detectable amount of photochemical or thermal debromination. The molecular structure of this universal polar solvent shows a trigonal pyramidal geometry about the central sulfur atom, which also contains a lone pair of electrons directed approximately to a tetrahedral apex, and is shown in Figure 4.

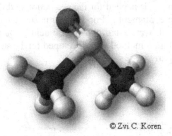

© Zvi C. Koren

Figure 4. Molecular structure of $(CH_3)_2SO$, the DMSO (dimethyl sulfoxide) molecule, the optimal solvent for the dissolution of the purple pigment.

Typically, a micro-sample weighing less than 50 μg (barely visible to the naked eye) is placed in a 2-mL glass vial and 100 – 400 μL of DMSO is added to it and placed in a dry-block heater previously set to 150 °C and heated for 5 minutes. This is then followed by microfiltration of the solution via centrifugation for 5 minutes by means of a centrifuge tube assembly containing a nylon filter of 0.2 – 0.45 μm pore size housed in a polypropylene body.

ANALYTIC METHOD DEVELOPMENT: HPLC OPTIMIZATION – SEPARATION/DETECTION OF ALL THE COLORANTS CONSTITUTING THE PURPLE PIGMENT

The common and abbreviated names for the possible dyes constituting molluskan pigments are listed in Table 1.

Table 1. Common and abbreviated names of the dyes.

Chemical group	Abbreviation	Common name
Isatinoids	IS	isatin
	4BIS	4-bromoisatin
	6BIS	6-bromoisatin
Indigoids	IND	indigo
	MBI	6-monobromoindigo
	DBI	6,6'-dibromoindigo
Indirubinoids	INR	indirubin
	6MBIR	6-monobromoindirubin
	6'MBIR	6'-monobromoindirubin
	DBIR	6,6'-dibromoindirubin

The method by which the components are separated so that they are extracted and washed out of a chromatographic column is known as "elution" and the mobile phase (the solvent) that performs this task is termed the "eluent". A suitable elution method – combining solvents, concentrations, and gradient times – is needed for the separation and detection of all the possible colorants constituting the purple pigment. For this purpose, a ternary solvent system was developed consisting of methanol, water, and phosphoric acid (5 % w/v, pH of 1.50 at 25 °C) [13, 23]. This linear gradient elution method was developed for 10 standard purple-related dyes, and its method is depicted in Figure 5.

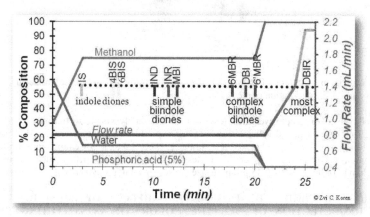

Figure 5. Gradient elution method used in the HPLC analyses showing the time-dependence of the composition of the ternary solvent system (methanol, water, and phosphoric acid) as well as the flow rate, and a schematic depiction of the separation of the ten colorants studied (© Zvi C. Koren).

This scheme produces good separation for all of the major dyes investigated, and the resulting chromatogram – a depiction of the chromatographic separation – is shown in Figure 6. This method optimizes the separation of these components in less than 30 minutes. The time that each component is retained in the separation column before eluting out is called the retention time, and labeled as t_R or R.T.

The advantage of this method over those previously published is that it produces a good separation of the smaller isatinoid molecules that may be present in the pigment, whereas other methods did not include this group in their separation scheme. Though isatin itself has been shown to be a truly minor component in the raw unprocessed purple pigment and in dyeings subsequently produced from these pigments [19, 24], however, the raw *B. brandaris* and *S. haemastoma* pigments analyzed by this method contain a major quantity of 6BIS, which may be useful in differentiating such pigments from the *H. trunculus* ones [19].

It should be noted, however, that the separation between DBI and 6'MBIR is not ideal, and in practice, the concentration of DBI is much greater than 6'MBIR and thus the former will often mask the latter's peak, as they nearly co-elute. Future research should be focused on producing a better separation of these two components.

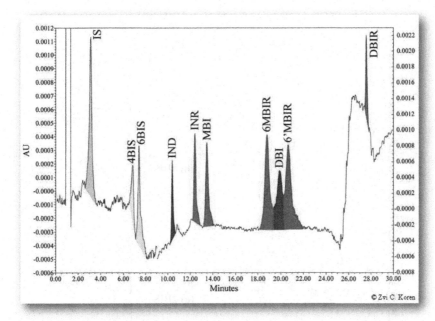

Figure 6. HPLC chromatogram showing the separation of the dyes, where the left vertical scale is for the first three isatinoids, and the right scale is for the other dyes; the abbreviations are explained in Table I (© Zvi C. Koren).

With this method, a spectrometric detection of the eluting components is made by the photodiode array (PDA) detector – also abbreviated as DAD (diode array detector). It produces a UV/Vis absorption spectrum for each dye detected. These are shown in Figures 7–9 for the three related chemical groups that constitute the purple pigment, the isatinoids, indigoids, and indirubinoids. Included in the figures are the molecular formulas and abbreviations of these dyes. From the respective spectra, it can be seen that the isatinoids maintain their color when dissolved producing yellowish solutions at about 415 nm. Similarly, the indirubinoids in DMSO-solution approximately retain their reddish-purple hue in the dissolved state as in the solid state, with an absorption wavelength at a maximum visible absorption, λ_{max}, of about 540 nm. While the navy-blue indigo produced blue DMSO-solutions, as expected, however, contrary to expectations, the violet MBI and purple DBI, both produce bluish solutions when dissolved. In fact, visually, in dilute solutions, all three indigoids produce similar blue-colored solutions, with λ_{max} values from about 600 – 615 nm.

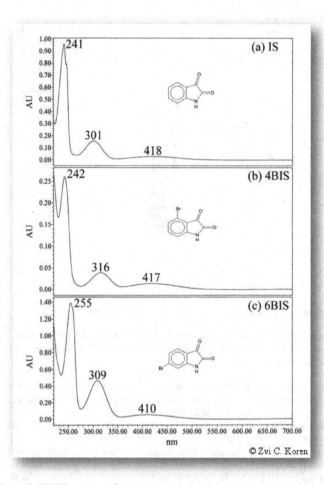

Figure 7. UV/Vis spectra of the isatinoids in DMSO solution (© Zvi C. Koren).

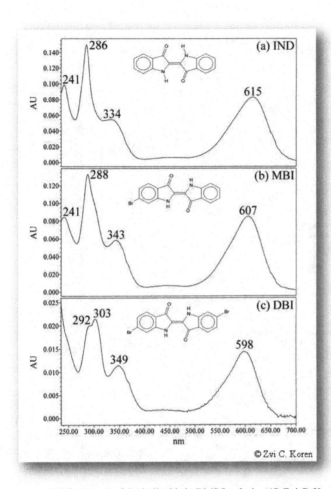

Figure 8. UV/Vis spectra of the indigoids in DMSO solution (© Zvi C. Koren).

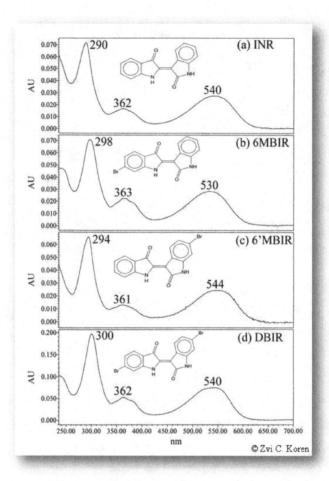

Figure 9. UV/Vis spectra of the indirubinoids in DMSO solution (© Zvi C. Koren).

The two properties produced from the HPLC-PDA methodology, the chromatographic retention time and the spectrometric UV/Vis absorption profile, provide a dual means of positive identification of each dye. Thus, by comparing the retention time and the spectrum of each peak in the chromatogram produced from a real sample with those two properties from a standard component, the identity of the unknown component can be identified with a high confidence level.

APPLICATION OF THE ANALYTIC HPLC METHOD TO THE CHROMATIC FINGERPRINTING OF MODERN *MURICIDAE* PIGMENTS

In order to determine the zoological provenance of purple archaeological pigments produced from *Muricidae* snails, it is important to identify and characterize the presence of all the detectable colorants produced from each zoological species. The unique "chromatic fingerprinting" is important in order to perform archaeo-malacological provenance determinations of ancient purple pigments and dyes. It is of course assumed that the living species available today were also in existence in antiquity.

The chromatographic multi-component characterizations of selected samples from modern *Muricidae* species, *H. trunculus, B. brandaris,* and *S. haemastoma,* are shown in Figure 10.

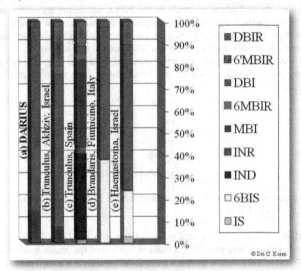

Figure 10. Bar chart showing the uncorrected dye compositions of five purple pigments expressed as percent integrated peak areas at 288 nm: (a) an archaeological pigment from a 2,500 year-old jar attributed to King Darius the Great, (b) modern *H. trunculus* snail from Akhziv, Israel, (c) modern *H. trunculus* from Spain, (d) modern *B. brandaris* from Fiumicino, Italy, and (e) modern *S. haemastoma* from Israel (© Zvi C. Koren).

The application of the analytical HPLC method to study the pigments from modern Muricidae snails has led to the following advances and discoveries made in this laboratory:

- First HPLC analysis of a raw unprocessed snail pigment [12];
- First HPLC identification of DBIR in a *H. trunculus* snail [12];
- First successful HPLC separation and detection of ten colorants that could constitute a molluskan purple pigment [12, 13, 19, 22]
- First HPLC detection of the following four dyes in *H. trunculus* snails: IS, 6BIS, 6MBIR, 6'MBIR [13, 22]

- First HPLC detection of 6BIS in *B. brandaris* and *S. haemastoma* [19]
- First *optimized* all-natural fermentation dye vat consisting of *H. trunculus* snails [18, 22]; a related natural process was also independently re-discovered by the late John Edmonds [25] and Inge Boesken-Kanold [26].
- First determination that a significant quantity of MBI in a purple pigment, modern or archaeological, indicates that the source of that colorant is *H. trunculus* [19].

ARCHAEOMETRIC ANALYSES OF RESIDUAL PURPLE PIGMENTS AND DYES

Tel Dor – 6th cent. BCE pigment at a Phoenician dyeing installation

Prior to the popularity and advances made by the HPLC technique and instrumentation, a widely used instrumental method was visible spectrophotometry whereby the sample is dissolved and a spectrum of light absorption is obtained at various wavelengths. While limited in its powers of discernment, it nevertheless was a simple instrumental tool and useful when carefully interpreted. Thus, a dark residual stain on a small piece of limestone (Figure 11) found in the conduit between two pits at Tel Dor, north-central Israel, dating from the 6th cent. BCE (Figure 12), was investigated.

© Zvi C. Koren

Figure 11. Small piece of an archaeological limestone with a dark stain found in a Tel Dor dyeing installation (© Zvi C. Koren).

The dark residual pigment was analyzed by dissolving it in hot DMF [27]. After about a minute, the solution's color turned pale blue, a color indicative of the presence of an indigoid in the residue. The color may be due to the presence of indigo alone, which would thus indicate that this site was a dyeing installation for producing blue-dyed textiles from a flora source, most probably from the leaves of the woad plant, *Isatis tinctoria*. Alternatively, if the solution's blue color was also a result of the presence of brominated indigoids (see above), which can only originate from fauna sources, then that would indicate that the vat dyeing installation was for the production of purple dyeings from mollusks. The results of the analyses are depicted in Figure 13 and show a spectrum for the dissolved archaeological residue with a visible wavelength at maximum absorption, λ_{max}, of 600 nm. Comparing this value with the comparable wavelength for indigo (613 nm) and with that for DBI (598 nm) shows that the now soiled residual pigment was much richer in the red-purple DBI than in indigo and hence the source for this pigment is in fact molluskan. The possible presence of MBI (absorbing at 607 nm) can also not be ignored. Hence, this installation served in fact for the dyeing of real purple textiles. The limestone found between the pits may have been used for the production of an alkaline environment, which is necessary for the reductive dissolution of the dye to its leuco form during the pre-dyeing stage.

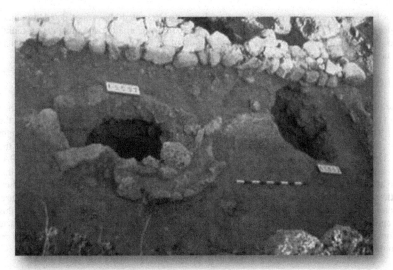

Figure 12. The two-pit installation connected via a conduit at Tel Dor (courtesy of the Eretz Israel Museum, Tel Aviv).

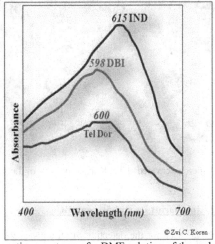

Figure 13. UV/Vis absorption spectrum of a DMF solution of the archaeological dark pigment from Tel Dor as compared with the spectra of indigo (IND) and 6,6'-dibromoindigo (DBI) (© Zvi C. Koren).

This type of visible spectrophotometric analysis was, as noted above, not detailed enough and hence the advent of the HPLC method provided the fine qualitative and quantitative tuning that was paramount for a complete determination of the dyes constituting archaeological pigments.

Tel Kabri – 7th cent. BCE purple-stained potsherd

The first HPLC method to study an archaeological pigment from a dyeing vat was published in 1995 [12]. The beautiful potsherd sample, excavated at the Phoenician site of Tel Kabri in the north of Israel, is depicted in Figure 14. For this analysis the detector was a variable wavelength detector, but measures the chromatogram at a fixed wavelength chosen at the beginning of the run. While this detector lacked the universality of the more modern photodiode array (PDA) detectors, it nevertheless provided accurate data for the most optimal wavelength, which in this case is about 600 nm. As previously noted, brominated indigoids and indigo have their absorption maxima near this wavelength. Since these analyses are destructive in nature, microsamples are needed and most of the time these analyses are for one time only. At this wavelength (600 nm), the existence of DBIR is still detected even though its wavelength at maximum absorption is about 540 nm. This was the first time that DBIR was found in an archaeological pigment.

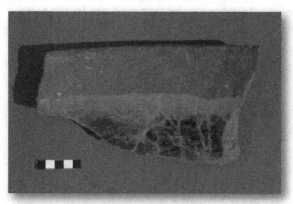

Figure 14. Purple-stained potsherd excavated at Tel Kabri (courtesy of the Tel Kabri Expedition, Tel Aviv University).

Purple-painted royal jar of King Darius I – mid-5th cent. BCE

The photodiode detector was used for the detection of an unusual royal object [19]. This unique marble jar with the name of King Darius carved on it contains sporadic purple stains on its exterior and even on the base (Figure 15). Upon closer inspection, it can be extrapolated that the entire object was painted with a purple pigment in a fresco-style technique. The mortar used was probably kaolinite and then painted afterwards with the purple pigment. The detailed chromatographic fingerprint (see Figure 10a) shows that this reddish purple pigment was obtained from a *H. trunculus* snail (or very similar species) that is rich in DBI and poor in IND (indigo). During the famous Persian Achaemenid Period, the kings controlled a vast empire with easy access to Mediterranean and other shores inhabited by *Muricidae* snails.

Figure 15. The King Darius I stone jar (courtesy of the Bible Lands Museum, Jerusalem).

The function and purpose of this object is not entirely certain, but it was probably a gift from the king to one in whom the king found favor. It is interesting to note a Biblical parallel here. In the sixth chapter of the Book (or Scroll) of Esther, in which according to tradition, the story unfolds in the royal court of one of the Achaemenid kings, probably Xerxes, it states:
"In this manner shall be done to the man whom the king desires to honor."

Herodian fabric – late 1st cent. BCE

One of the more professionally gratifying and personally emotional discoveries made in this laboratory was the determination that a miniscule piece of fabric, measuring about 2 x 4 mm

(Figure 16), found in a garbage dump at the top of the palatial fortress of Masada in the Judean Desert belonged to the royal purple mantle of King Herod [24, 28, 29]. This king was probably the most colorful of kingly characters reigning in Judea in ancient Israel. He built a number of palaces and other building projects, some say even megalomaniac in size [30], so much so that he probably had an "Edifice Complex". But probably his most outstanding landmark was the refurbishment of the Second Temple in Jerusalem, of which the famous external Western Wall still survives with its Herodian-stone grandeur.

Figure 16. The miniscule purple fabric found at Masada (© Zvi C. Koren).

The chromatographic result of an HPLC analysis on a small dyed yarn from that weave (Figure 17) shows the trademark of a dyeing produced from a molluskan source. Comparing this dye composition with that of other snails (see Figure 10), and especially based on the fact that it has a significant quantity of MBI, it is clear that the marine zoological source of this pigment is the *H. trunculus* snail (or some other species that is nearly indistinguishable from it). This pigment is rich in DBI, which is responsible for its reddish-purple hue. This then is also the color of the Biblical *Argaman* dye mentioned above, the first such dyeing found in ancient Israel.

Recently, the author discovered the other biblical dye, *Tekhelet*, a blue-purple color derived from an indigo-rich pigment also produced from the *H. trunculus* snail on a woolen dyeing from Masada, and reported in The New York Times [31]. A detailed scientific report on this unprecedented find will be forthcoming soon.

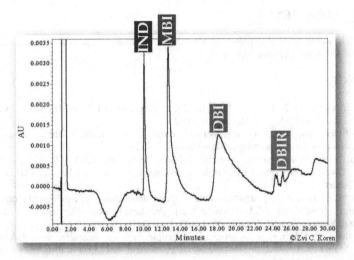

Figure 17. Chromatogram at 600 nm of the dye extracted via DMSO from the purple yarn from the Herodian fabric (© Zvi C. Koren).

CONCLUSIONS

The analytical study of purple molluskan pigments and dyes on archaeological artifacts from more than two and a half millennia has shown major advancements, especially within the last two decades. The advances and discoveries in this area made in this laboratory can be summarized as follows:

- Decipherment of the method by which the ancients performed purple-dyeing by natural means [18];
- First HPLC analysis of a raw unprocessed purple archaeological snail pigment [12];
- First HPLC identification of DBIR in a molluskan archaeological pigment [12];
- First discovery of the purple pigment as the sole paint pigment on a royal objet d'art from King Darius I [19];
- First discovery of the royal purple mantle of King Herod I – the Biblical color Argaman [24, 28, 29].

Now that the color of the reddish-purple Biblical *Argaman* or the Septuagint's πορφυραν dye has been discovered, the mystery surrounding the color of *Tekhelet* or υακινθον, the other "purple", will be divulged soon.

ACKNOWLEDGEMENTS

The author would like to express his sincere appreciation to the Sidney and Mildred Edelstein Foundation for support of this work.

REFERENCES

1. Aristotle. *History of animals*. Peck A L (transl.), Loeb Classical Library, Book V, Harvard University Press, Cambridge, Massachusetts, 2002.
2. Pliny the Elder. *Natural history*. Rackham H (transl), Loeb Classical Library, Book IX, Harvard University Press, Cambridge, Massachusetts, 1952.
3. D. Cardon, *Le monde des teintures naturelles*. Éditions Belin, Paris, 2003. 418.
4. D. Cardon, *Natural dyes – sources, tradition, technology and science*. Archetype Publications, London, chapter 11, 2007.
5. R. Haubrichs, L'étude de la pourpre: histoire d'une couleur, chimie et expérimentations. In: *Conchiglie e Archeologia*. M.A. Borrello (ed.), Preistoria Alpina 40, Suppl. 1. Museo Tridentino di Scienze Naturali, Trento, 2004, 133.
6. R. Haubrichs, Natural history and iconography of purple shells. in: *Indirubin, the red shade of indigo*. L. Meijer, N. Guyard, L. Skaltsounis, G. Eisenbrand, eds., Life in Progress Editions, Roscoff, chapter 6, 2006, 55.
7. C.J. Cooksey, *Molecules* 6 (2001) 736.
8. C.J. Cooksey, Marine indirubins, in: *Indirubin, the red shade of indigo*. L. Meijer, N. Guyard, L. Skaltsounis, G. Eisenbrand, eds., Life in Progress Editions, Roscoff, chapter 3, 2001, p 23.
9. J. Wouters, A. Verhecken, *J. Soc. Dyers Colour* 107 (1991) 266.
10. J. Wouters, *Dyes Hist. Archaeol.* 10 (1992) 17.
11. Z.C. Koren, *J Soc Dyers Colour* 110 (1994) 273.
12. Z.C. Koren, *Isr. J. Chem.* 35 (1995) 117.
13. Z.C. Koren, HPLC-PDA analysis of brominated indirubinoid, indigoid, and isatinoid dyes. in: *Indirubin, the red shade of indigo*. L. Meijer, N. Guyard, L. Skaltsounis, G. Eisenbrand, eds., Life in Progress Editions, Roscoff, chapter 5, 2006, p. 45.
14. I. Karapanagiotis, *Amer. Lab.* 38 (2006) 36.
15. I.Karapanagiotis, V. de Villemereuil, P. Magiatis, P. Polychronopoulos, K. Vougogiannopoulou, A.L. Skaltsounis, *J. Liq. Chrom. Rel. Tech.* 29 (2006) 1491.
16. S. Sotiropoulou, I. Karapanagiotis, (2006) Conchylian purple investigation in prehistoric wall paintings of the Aegean area. in: *Indirubin, the red shade of indigo*. L. Meijer, N. Guyard, L. Skaltsounis, G. Eisenbrand, eds., Life in Progress Editions, Roscoff, chapter 7, 2006, 71.
17. E. Spanier, N. Karmon, Muricid snails and the ancient dye industries, in: *The royal purple and the biblical blue: Argaman and tekhelet*. The study of Chief Rabbi Dr. Isaac Herzog on the dye industries in ancient Israel and recent scientific contributions. E. Spanier E, Keter, Jerusalem, 1987, 179.
18. Z.C. Koren, *Dyes Hist. Archaeol.* 20 (2005) 136.
19. Z.C. Koren, *Microchim. Acta* 162 (2008) 381.
20. Z.C. Koren, *Where Have All the Safflower Reds Gone?* 17th Meeting of Dyes in History and Archaeology, National Maritime Museum, Greenwich, UK November, 1998.
21. Z.C. Koren, *Dyes Hist. Archaeol.* 16/17 (2001) 158.

22. Z.C. Koren, *The Purple Question Reinvestigated: Just What is Really in That Purple Pigment?* , Abstracts of the 20th Meeting of Dyes in History and Archaeology, Department of Conservation Research, Netherlands Institute for Cultural Heritage, Amsterdam, Holland; November, (2001) 10.

23. Z.C. Koren, *Dyes Hist. Archaeol.* 21 (2008) 26.

24. Z.C. Koren, *Non-Destructive vs. Microchemical Analyses: The Case of Dyes and Pigments,* Proceedings of ART2008, 9th International conference, non-destructive investigations and microanalysis for the diagnostics and conservation of cultural and environmental heritage, May 25-30, Jerusalem, Israel, 2008,371.

25. J. Edmonds, *The mystery of imperial purple dye* [cover title: *Tyrian or Imperial Purple Dye*], Historic Dye Series no. 7, author's self-publication, Little Chalfont, UK. 2000.

26. I.B. Kanold, *Dyes Hist. Archaeol.* 20 (2005) 150.

27. Z.C. Koren, *Dyes Hist. Archaeol.* 11 (1993) 25.

28. Z.C. Koren, The unprecedented discovery of the Royal Purple dye on the two thousand year-old royal Masada textile. American Institute of Conservation, *The Textile Specialty Group Postprints* 7 (1997) 23.

29. Z.C. Koren, Color my world: a personal scientific odyssey into the art of ancient dyes, in: *For the sake of humanity: essays in honour of Clemens Nathan. Martinus Nijhoff – Brill*, A. Stephens, R. Walden R (eds.), Leiden, 2006, 155.

30. N. Kokkinos, *The Herodian Dynasty*. Spink, London, 2010.

31. http://www.nytimes.com/2011/02/28/world/middleeast/28blue.html?emc=eta1. (accessed Feb. 28, 2011).

Mater. Res. Soc. Symp. Proc. Vol. 1374 © 2012 Materials Research Society
DOI: 10.1557/opl.2012.1377

Characterization of a Natural Dye by Spectroscopic and Chromatographic Techniques

Y. Espinosa-Morales[1], J. Reyes[1], B. Hermosín[2], J. A. Azamar-Barrios[3]

[1]Centro de Investigación en Corrosión, Universidad Autónoma de Campeche. Avda. Agustín Melgar s/n entre Juan de la Barrera y Calle 20. Colonia Lindavista, San Francisco de Campeche, México. e-mail: javreyes@uacam.mx

[2]Instituto de Recursos Naturales y Agrobiología de Sevilla (IRNAS-CSIC), Spain.

[3]CINVESTAV-Unidad Mérida, México.

ABSTRACT

Natural dyes have been extracted from both plants and animal to give color to textiles and handicrafts. This is the case of purple dye extracted from *Justicia spicigera Schldt*, an acanthaceae used as a color source since pre-Hispanic period in the Mayan area of Mexico and Central America. Spectroscopic (UV-Vis and FT-IR) and chromatographic (PY-GC/MS) techniques were employed in order to characterize some of their chemical properties. UV-VIS absorption spectra indicates a λ_{max} peak at 581 nm, value associated to anthocyanins group under bathochromic effect. On the other hand, a structural characterization realized by FT-IR and Py-GC/MS indicated the presence of polar hydroxibenzoic acids and phenolic compounds which are characteristics of the molecular structure of anthocyanins.

INTRODUCTION

Mesoamerican culture had characterized by the use of variety of dyes extracted from natural sources like minerals, plants and animals. Dyes were employed to get color to textiles, ceramics, mural and also corporal paint for native population of America, where color had several ceremonial and quotidian contexts [1, 2].

In Mexico, it is highlighted the blue color extracted from añil (*Indogofera Tictoria* Linn). The exotic blue Maya for example, was made by mixig the extract of añil with paligoskite clay, getting high stability along the time [3].

In the Maya region of Mexico and Central America, people learned to value the use of natural sources of dyes from plants like Achiote (*Bixa Orellana*), Indigo, Palo de Campeche (*Haematoxylum campechianun*), Kanté (*Diphysa robinoides* Mill.) and Chak Lool (*Justicia spicigera Schltd*) among others [1, 3]. The Spanish colonizers gave great importance to the discovery of the native species and variety of plants used as a color sources in America [4, 5].

The leaves of Chak Lool plant have been used to extract a purple dye traditionally employed for artisans from the Camino Real Region in Campeche (México), to dye textiles and handicraft like the guano palm from they get the jipi hats, bags and carpets (Figure 1a and 1b). Nowadays, there is not enough information about its chemical structure and properties. Nevertheless, it is know that purple colors produced by plants are related to the flavonoids group called anthocianins [6].

More studies are necessary in order to include the chemical properties of the Chak Lool dye extract in a Mexican color data base [7]. For this purpose, actually exist diverse spectroscopic and chromatographic techniques widely used to characterize the properties of natural compounds like colorants.

Figure 1. (a) General appearance of the leaves of *Justicia Spicigera Schltdl* (Chak Lool plant). (b) Artisanal purple dye extraction from Chak Lool plant.

Analytical pyrolysis coupled to gas chromatography/mass spectrometry (PY-GC/MS), is a chromatographic variant used to analyze organic matter in short time and minimum sample preparation [8]. It allows the study of biopolymers or geopolimers, and naturals products like superior plant metabolites, from the analysis and interpretation of its pyrolysis products [9-14].

On the other hand, Ultraviolet/Visible absorption (UV/VIS) and Fourier Transformed–Infrared (FT-IR) spectroscopies can be used as qualitative tool to identify and characterize molecular species or some of their properties like molecular structures and characteristic absorption spectra that at least constitute their characteristic fingerprints.

This contribution shows the results of the use of these techniques in order to analyze the purple dye extracted from leaves of Chak Lool plants. The study allowed the generation of its chemical characteristic profiles and the identification of some structures probable associated to its purple color.

MATERIALS AND METHODS

Obtaining the dye extract

5 g of fresh leaves from Chack Lool plant was crushed in a glass mortar and extracted with 100 ml of the aqueous pure distilled water during one hour in an ultrasound bath (Cole Parmer). After that, the purple dye extract was filtered across 21 mm glass fiber membrane (GF/A, Whatman), then kept at 3°C protected from the direct light (Figure 2a).

On the other hand, 5 g of leaves were dried in a convention oven at 45 °C during 24 hours and extracted following the procedures previously described. In both cases, 15 ml of each extract was lyophilized in a Labconco freeze-dry system in order to realize FTIR and Py-GC/MS analysis (Figure 2b).

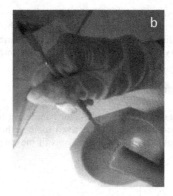

Figure 2. (a) Aqueous dye extract of obtained from leaves of Chack Lool plant. (b) Lyophilized dye extract.

UV-Vis spectroscopy analysis

1 ml of the total aqueous dye extract was diluted in 100 ml of pure distilled water. An aliquot was introduced in a quartz cell (1 cm pathway) and analyzed in a Jenway UV/Vis spectrophotometer. A scan from 300 to 600 nm was performed in order to generate the characteristic absorption spectra of the sample.

FT-IR spectroscopy analysis

5 mg of lyophilized dye extract were mixed thoroughly with 195 mg of potassium bromide (KBr) until homogenized in an agate mortar. The mixture was placed into the sample compartment of a Smart Collector Diffuse Reflectance Infrared Fourier Transform Spectroscopy (DRIFTS) attachment of the spectrometer, which was continuously purged with dry air. Then, analyzed in a Thermo Nicolet Nexus 670 FT-IR spectrophotometer equipped with a DTGS KBr detector and a purge gas generator; at a spectral resolution and wave number precision of 0.09 and 0.01 cm^{-1} respectively. For each spectrum, 32 scans were used.

PY-GC/MS analysis

A few mg of lyophilized dye extract and grinded dry leaves from Chak Lool were deposited in a separate Curie-point small hollow ferromagnetic cylinder (temperature 300°C), dried and wetted with 5 µl of methanolic solution of tetramethyl-amonium hydroxide (25 % w/w). The cylinder was slightly dried with a N_2 flow and immediately inserted in the pyrolyser.

The analysis was performed in a Fisons instrument GC 8000 Top/VOYAGER, using a 30 m x 0.25 mm ZB 5ms column (film thickness 0.25 µm), coupled to Fischer 0316 Curie-point pyrolyser. The GC oven was programmed from 50°C to 280°C, at a rate of 5°C min^{-1}. This temperature was held for 30 min and then to 310°C at 20°C min^{-1}, where the final temperature was held for 2 min.

The mass detector operated at ionization energy of 70 eve, with a scan rate of 0.9 sec. The mass spectra identification was carried on by using own spectra collection and the NIST electronic library from the Unites State of America. In the analytical procedure used in this work, the carboxylic acids were recovered as the corresponding methyl esters and the hydroxyls as methoxyls.

DISCUSSION

The characteristic purple color manifested by the dye extract could be the consequence of the presence of compounds derived room anthocyanins group. Figure 3 shows the absorption spectra generated during the analysis of the aqueous dye extract obtained from leaves of Chak Lool plant.

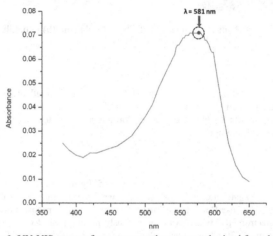

Figure 3. UV-VIS spectra from aqueous dye extract obtained from leaves of Chak Lool plant.

A maximum absorbance peak (λ_{max}) was detected at 581 nm. This value does not match with those corresponding to anthocyanins like petunidin ($\lambda_{max} = 543$ nm), malvidin ($\lambda_{max} = 542$ nm) or delphinidin ($\lambda_{max} = 546$ nm). This fact suggests a bathochromic effects probably caused by co-pigmentation mechanisms. It is consequence of a typical anthocyanins intermolecular association within flavonoids like tannins and catechins or glucocydic groups which can act like co-pigments.

Co-pigmentation mechanisms have been studied by several authors [15-18]. They reported that the increasing in cafeic acid concentration shift the λ_{max} to highest value and also increase the possibility of hyperchromic effects in anthocyanins at pH value between 3.0 and 4.0.

In 1915, Willstates and Mieg [19] related the red-purple color of *Delphinium ajacis*, with delphinidin molecule. It have been observed that when delphinidin suffer co-pigmentation, its λ_{max} is shifted until closer values to that observed for Chak Lool aqueous extract ($\lambda_{max} = 581$ nm).

They reported that an increasing in cafeic acid concentration shifts the λ_{max} to highest value and also increase the possibility of hyperchromic effects in anthocyanins at pH value between 3.0 and 4.0. Other reports indicate that during the formation of cafeic acid/delphinidin complex, λ_{max} value is shifted until 585 nm [16, 20].

On the other hand, Schreiber et al. [18], observed that the complexation of delphinidin with metallic ions like Al^{3+} displaced λ_{max} even 589 nm. Anthocyanins are water soluble non-nitrogenous phenolic compounds.

This fact suggests the apparition of a bathochromic effects probably caused by co-pigmentation mechanisms. Co-pigmentation mechanisms have been studied by several authors [15-18].

They are a cationic form of flavil group [21]. Its structure is formed by two aromatic groups (Figure 4): the benzopiril group (A and C ring) and the phenolic group (B ring). Those compounds may have H, -OH or -OCH$_3$ substituent groups. Their characteristic color depends strongly from the number of substituents at the B ring (Table 1) [22].

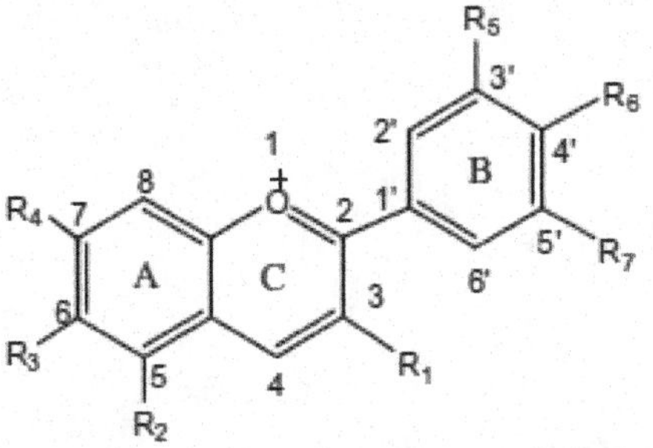

Figure 4. General structure of anthocyanins.

Table 1. Typical B ring substituent at the flavil group.

Anthocyanins	R_5	R_7
Pelargonidin	-H	-H
Cyanidin	-OH	-H
Delphinidin	-OH	-OH
Petunidin	-OH	-OCH3
Malvidin	-OCH	-OCH3
Peonidin	-OCH3	-H

In this order, FT-IR spectra in figure 5 of lyophilized extract shows a strong signal corresponding to the characteristic C-C=C aromatic group vibrational frequency (1618-1645 cm⁻¹). At 795-825 cm⁻¹, a weak absorption band from the aromatic C-H bond out the plane flexion movement was observed [23, 24]. Between 1405 and 1416 cm⁻¹ the flexion movements corresponding to the phenolic -OH group was presented. A wide band from OH group was manifested in the range of 3222-3310 cm⁻¹, while the typical vibrational frequency from C-O bond stretching movement was observed at 1233-1224 cm⁻¹. It was an indicative of the presence of phenolic compounds in the molecular structure of the dye.

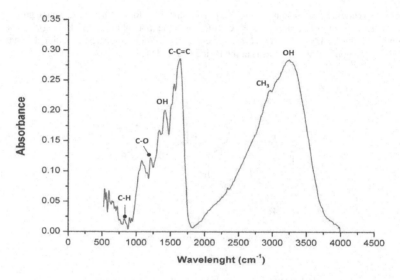

Figure 5. FT-IR spectra from lyophilized purple dye extract obtined from leaves of Chak Lool plant and associated funtional group.

On the other hand, the asymmetric movements from -CH₃ group appeared in the range of 2922 to 2973 cm⁻¹. That structure could be associated to the methoxyl- group (-OCH3), which is usually related with purple color when it is a substituent in the phenolic B ring of the flavil group of anthocyanins.

Table 2, lists the major compound identified during the PY-GC/MS analysis from both the the drye leaves and the lyophilized purple dye extracts (from fresh and dry leaves) of the Chak Lool plant.

Figure 6 shows the Total Ion Chromatogram (TIC) from their respective pyrolysates. According the results, the predominant identified structures correspond to phenolic derivatives like methoxybenzoic acids (group A), phenolic acids (group B), n-carboxilic acids (group C). Also derivatives from cellulose (group D), alkaloids (group E) and terpenoid hydrocarbons (group H) could be observed. Figure 7 shows the major structures identified.

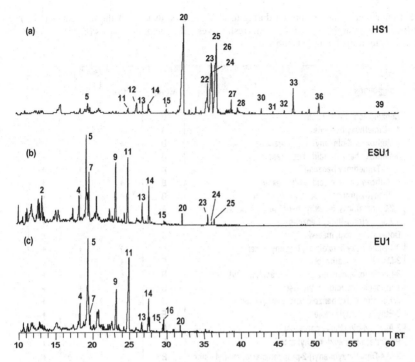

Figure 6. Total ion chromatogram (TIC) from: (a) dry leaves plant (HS1), (b) and the lyophilized purple from the dry leaves (ESU1) and (c) fresh leaves (EU1) of Chak Lool. RT, Retention Time (minutes).

The TIC corresponding to HS1 sample (Figure 6 a), shows the methyl ester of hexadecanoic acid (peak 20) and the isomeric octadecatrienoic acid methyl ester (peaks 22 to 26) like the major compounds. In this sample, a greater abundance of *n*-carboxilic acids in the range of C_{12} to C_{30} is noticeable. *n*-Carboxilic acids are derived from cuticular wax of superior plants [13].

It was noticeable the relative high intensity of the peak 33, that corresponded to the acyclic isoprenoid hydrocarbon squalene (2,5, 10, 15, 19, 23-hexametyl tetracosane), that is precursor of polycyclic triterpenoids in animals and plants. This compound was isolated for first time in shark liver oil [25]. Recently, its presence was reported in chromatographic profiles of melastomataceae plants [26]. Phenolic and methoxybenzoic groups appear only as minor compounds within the furanone 4-methoxy-2,5-dimethyl-3(2H)-furanone (peak 3) and the alkaloid 3-methoxycarbazol (peak 12).

Table 2. Major compounds identified during the PY-GC/MS analysis of the dry leaves (HS1) and the lyophilized purple dry extract from fresh leaves (EU1) and, dry leaves (ESU1) of Chak Lool Plant. (+) identified, (-) not identified.

Peak	Compounds	Group	HS1	ESU1	EU1
1	Benzoic acid, methyl ester	B	-	+	-
2	1,4-Dimethoxy benzene	A	+	+	+
3	4-Methoxy-2,5-dimethyl-3(2H)-furanone	D	+	-	-
4	3-Methoxy benzoic acid, methyl ester	B	+	+	+
5	1,2,4-Trimethoxy benzene	A	+	+	+
6	4-Methoxy benzoic acid, methyl ester	B	+	+	+
7	3-Phenyl-2-propenoic acid, methyl ester	B	+	+	+
8	1,2,3-Trimethoxy-5-methyl-benzene	A	-	+	+
9	1,2,3,4-Tetramethoxy benzene	A	-	+	+
10	Dodecanoic acid, methyl ester	C	-	+	+
11	3,4-Dimethoxy benzoic acid, methyl ester	B	+	+	+
12	3-Methoxy-9H-carbazole	E	+	-	-
13	3-(4-methoxyphenyl)-2-propenoic acid, methyl ester	B	+	+	+
14	Tetradecanoic acid, methyl ester	C	+	+	+
15	3,4,5-Trimethoxy benzoic acid, methyl ester	B	+	+	+
16	3-Ethyl-2-phenyl indole	E	-	-	+
17	3-Allyl-1,4-dimethoxy carbazole	E	-	-	+
18	Pentadecanoic acid, methyl ester	C	+	+	+
19	3-(3,4-Dimethoxyphenyl)-2-propenoic acid, methyl ester	B	-	-	+
20	Hexadecanoic acid, methyl ester	C	+	+	+
21	Heptadecanoic acid, methyl ester	C	+	-	-
22	Octadecatrienoic acid, methyl ester	C	+	-	-
23	Octadecatrienoic acid, methyl ester	C	+	-	-
24	Octadecanoic acid, methyl ester	C	+	+	+
25	Octadecatrienoic acid, methyl ester	C	+	-	-
26	Octadecadienoic acid, methyl ester	C	+	-	-
27	9,12,15-Octadecatrienoic acid, methyl ester	C	+	-	-
28	Eicosanoic acid, methyl ester	C	+	-	-
29	Heneicosanoic acid, methyl ester	C	+	-	-
30	Docosanoic acid, methyl ester	C	+	-	-
31	Tricosanoic acid, methyl ester	C	+	-	-
32	Tetracosanoic acid, methyl ester	C	+	-	-
33	Squalene	H	+	-	-
34	Pentacosanoic acid, methyl ester	C	+	-	-
35	Hexacosanoic acid, methyl ester	C	+	-	-
36	Heptacosanoic acid, methyl ester	C	+	-	-

Figures 6b and 6c show the TIC from the lyophilized purple dye extract (ESU1) and fresh leaves (EU1) from Chak Lool plant. Figure 7 shows the most important structures of phenolic and carboxylic acids peak compounds identified in the pyrolysates. In both cases, aqueous extraction allowed the preferential separation of the most polar compounds present in leaves, like phenolic compounds and its methoxy- and carboxylic derivatives.

The last one coincides with FT-IR spectroscopic characterization realized to EU1 samples. n-Carboxylic acid are only separated on trace levels under C_{20}, while compounds like squalene or furanones were not extracted. Majority compounds included 1,2,3-trimethoxybenzene (peak 5), 1,2,3,4-tetramethoxybenzene (peak 9) and 2,4-dimethoxybenzoic acid methyl ester (peak 11). Other compounds also separated were 3-phenyl-2-propenoic acid methyl ester and 3-(3,4-methoxyphenyl)-2-propenoic acid methyl ester (cafeic acid) corresponding to structures labeled as peaks 7, 13 and 19 respectively (Figure 7).

It is important to highlight that ESU1 such as EU1 showed the same chromatographic profile, because of that, molecular structure content in both samples could be considered the dye fingerprints.

Diverse authors have related these structures with flavonoid compounds constituents of anthocyanins like cyanidin, pelargonidin or peonidin [9, 21,27]. Other authors also reported their presence in tannins and lignins [28, 29]. On the other hand, the identification of cafeic acid (structure of the peak 19) reinforce the idea of the probably apparition of co-pigmentation. That leads the shift of anthocyanins color associated λ_{max} until higher values such as was observed in the absorption spectra of the dye.

CONCLUSIONS

UV-Vis spectroscopy allowed the generation of the characteristic absorption spectra from the aqueous purple dye extract from leaves of Chak Lool plant. A λ_{max} at 581 nm was established. This value seems to be consequence of bathochromic effect caused by a probably co-pigmentation mechanisms induced by the presence of compounds like cafeic acid.

FT-IR analysis allowed the identification of functional groups that correspond to phenolic compounds associated to $-CH_3$, -H and $-OCH_3$ group. They are characteristic substituent of the anthocyanins group.

A characteristic PY-GC/MS pattern was obtained from the lyophilized extract of Chak Lool plant. The analysis of the pyrolisates revealed the presence of polar compounds like hydroxibenzoic acids and phenolic compounds closely related to the structure of anthocyanins. They are probably the responsible of the characteristic purple color observed in the aqueous dye extract obtained from the leaves of Chak Lool plant. Finally the chemical structures identified the analysis of the purple dye extract obtained from leaves of Chak Lool Lool, could be considered its Fingerprint.

ACKNOWLEDGMENTS

This contribution was possible thanks to the support of the project "Extraction of natural dyes with potential application to restoration of artistic and historic works" from Universidad Autonoma de Campeche.

57

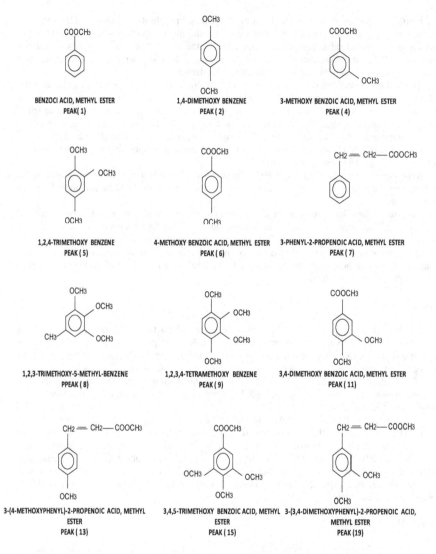

Figure 7. Structure of phenolic compound and their carboxylic acid identified in the lyophilized purple dye extract from dried and fresh leaves of Chack Lool plant.

REFERENCES

1. M. Ivic, M. Berguer. in: *Ciencia y técnica maya*. M. Ivic de Monterroso, I. Azurdia Bravo, Ed., 2008, 101-133.
2. D. Magaloni. *Arqueología Mexicana* **93**, (2008) 46-50.
3. D. Magaloni, in *La pintura mural prehispánica en México: Área Maya*. B. de la Fuente Ed., Instituto de Investigaciones Estéticas UNAM, Mexico, 2001, 155-198.
4. H. Cabezas. *Unidad geográfica cultural: flora mesoamericana*. Universidad Mesoamericana, Guatemala, 2005.
5. O. Lock, in: *Colorantes naturales*. Fondo Editorial. Pontificia Universidad Católica de Perú, Lima, 1997.
6. Y. Espinosa-Morales. Dissertation thesis. College of Chemistry and Biological Sciences. Universidad Autonoma de Campeche, Campeche, 2010.
7. E. Arroyo , T. Falcón, E. Hernández, S. Zetina, A. Nieto, J.L. Ruvalcaba. XVI *Century Colonial Panel Paintings from New Spain: Material Reference Standards and Non Destructive Analysis for Mexican Retablos*. 9th International Conference on NDT of Art, ART2008, Jerusalem, Israel (2008).
8. C. Saíz-Jimenez, K. Haider, H. L. C. Meuzelaar. *Geoderm.* 22 (1979) 25-37.
9. G. Lanzarini, L. Morselli, G.P. Pifferi, A.G. Giumanin. *J. Chrom.* 130 (1997) 261-266.
10. W. J. Irwin J. *Anal. Appl. Py.* 1 (1979) 3-25.
11. S. Ghosal, S Banerjee, D. Jaiswal. *Phytochem.* 19 (1980) 332-334.
12. C. Saiz-Jimenez, B. Hermosin, J. J. Ortega-Calvo. *Int. J. Environ. Anal. Chem.* 56 (1994) 63-71.
13. J. Reyes, Doctoral thesis. School of Chemistry. Universidad de Sevilla, Sevilla, 2004, 21-23.
14. F. Shadkami, R. J. Helleur, R. M. Cox, *J. Chem. Ecol.* 33 (2007) 1467-1476.
15. S. Asen, R.N Sterwart., K. H Norris. *Phytochem.* 11(3) (1972) 1139-1144.
16. E. Gris, A. Ferrari, L. Falcao, M. Bordignon, *Food Chem.* (2007) 1289-1296.
17. P. A. Kanakis, M. G. Tarantilis, S Poliss, H. A Diamantoglou, Tajmir-Riahi. *J. Mol. Struc.* 798 (2006) 69-74.
18. H. Schreiber, A. M. Swink, T. D Godsey. *J. Inorg. Biochem.* 104 (2010)732-739.
19. R. Willstater, W. Mieg Liebig. *Ann Chem.* 408 (1915) 327.
20. K. Yoshimata, K. Hayashi. Biol. Mag. Tokio. 87 (1974) 33-40.
21. I. Guerrero, E. Lopez, R. Armenta. Pigmentos. in: *Quimica de los Alimentos*. Salvador Badui Ed. Fourth Edition. volume II, Pearson Educacion, México, 2006, 401-439.
22. R. Myjavcová, P. Marhol, V. Kren, V. Simanek, J. Ulrichová, I. Palíková, B. Papouskova, K. Lemr, P. Bedñar, *J. of Chrom* A 1217 (2010) 7932–7941.
23. A.K. Satapathy, G. Gunasekaran, S.C. Sahoo, K. Amit, P.V. Rodrigues, *Corros. Sci.* 51 (2009) 2848-2856.
24. O. Torres, G. Santafé, A. Angulo, H. Villa, J. Zuluaga, M. Doria. *Sci. Tech.* XIII 3 (2007) 333-336.
25. R. P. Phillip, *Applications and spectra*. Elsevier, Amsterdam, 1985, 88-91.
26. J. H. Isaza, L. M. Orozco, L. M. Zuleta, D. A. Rivera, L. J. Tapias, L. A. Veloza, L. S. Ramirez. *Sci. Tech.* XIII 33 (2007) 359-362.
27. Z. Füzfai, I. Molnár-Perl. *J. Chrom.* A 1149 (2007) 88-101.
28. C. G.Nolte, J. J. Schauer, G. R. Cass, , B. R. T. Smoneit. *Environ. Sci. Tech.* 35 (2001)1912-1919.
29. M. P.Fernández, P. A.Watson, C. Breuil, J. Chrom. A. 922 (2001) 225-233.

Mater. Res. Soc. Symp. Proc. Vol. 1374 © 2012 Materials Research Society
DOI: 10.1557/opl.2012.1378

The Influence of Glass in the Color of Red Lakes Layers in Oil Painting: A Case Study in a Pictorial Series Attributed to Murillo Located in Guadalajara, Mexico

Elsa Arroyo[1], Adriana Cruz Lara[2], Manuel Espinosa[3], José Luis Ruvalcaba[4], Sandra Zetina[1], Elsa Hernández[1], Shannon Taylor[4]

[1] Laboratorio de Diagnóstico de Obras de Arte, Instituto de Investigaciones Estéticas, Universidad Nacional Autónoma de México, Circuito Mario de la Cueva s/n, Ciudad Universitaria, Mexico DF 04510, Mexico. e-mail: elsa2001@gmail.com
[2] Posgrado en Historia del Arte, Facultad de Filosofía y Letras, Universidad Nacional Autónoma de México, México DF, Mexico. e-mail: acruzlara2000@yahoo.com.mx
[3] Instituto Nacional de Investigaciones Nucleares, Carr. México-Toluca s/n La Marquesa, Ocoyoacac, 52750, Mexico. e-mail: meep@nuclear.inin.mx
[4] Instituto de Física, Universidad Nacional Autónoma de México. Mexico DF, Mexico. e-mail: sil@fisica.unam.mx

ABSTRACT

This paper discuss the presence of powdered glass and quartz integrated in the red lake layers of two paintings attributed to the Sevillian painter Bartolome Esteban Murillo that belongs to the Guadalajara's Regional Museum's art collection. A laboratory experimental reproduction of the Sevillian painting technique was made using three different lakes (cochineal, madder lake and brazilwood) mixed with four varieties of glass to explore the optical properties and the influence of the transparent and translucent aggregates into the oil paint layers. The experimental reproductions were analyzed using ultraviolet-visible spectroscopy, optical and fluorescence microscopy and scanning electron microscopy (SEM-EDX). A comparison between the originals and the reproduced red lakes layers was carried out to understand the artistic process of Murillo's color application. Preliminary results suggest that glass was not used as a siccative agent as the historical treatises mentioned but mainly as an additive to increase brightness, thickness and color saturation of the red lake layers related to the artist's intention.

INTRODUCTION AND PREVIOUS STUDIES OF PAINTINGS

A pictorial series with the iconography of the life of Saint Francis is located at Museo Regional de Guadalajara, Mexico. There are eleven big size oils on canvas, related by means of their iconography. Since 1876, on the basis of stylistic and formal similarities this cycle has been attributed to Bartolome Esteban Murillo, a capital figure of the 17th century Spanish painting but their origin is still unclear [1].

It is commonly admitted general resemblances with the style of the paintings produced in the Seville's workshops during that time, but recently, some researchers have argued a local New Spain's manufacture [2]. In order to discuss the art historian attribution, and to study the problems around the provenance and date of the Guadalajara's series, a technical examination of two paintings was performed. The paintings selected were *San Francisco resucita a un niño (St Francis resurrects a child)* and *La estigmatización de San Francisco (St Francis receiving the stigmata)* both are different in format size, paint handling and process of color application [3]. This paper is focused in the technical study of one of them: the narrative scene that represents the miracle when Saint Francis resurrects a child.

Figure 1. Bartolome E. Murillo (atrib.), *San Francisco resucita a un niño,* Oil on canvas, 320 x 410 cm, Museo Regional de Guadalajara, Mexico. Photography: Eumelia Hernández, 2010. LDOA-IIE, UNAM.

The dress of the noble woman is depicted as if it was made of soft fabric, it looks like silk and it has a light-pink color, with red and purple shadows and pinkish-yellow lights (Figure 3). The cross sections revealed a simple stratigraphy, with two or three layers overlapped, mixed with great amount of oil medium. The inner layer is lead white basically. Its function is to create a base-tone that hides the ground color, a brownish ocher surface. The paint layer that builds the volume of the pink dress was achieved by a sequence of mixtures applied simultaneously by brush, so the brushstrokes are merged in a chaotic manner. The red paint is translucent, a mixture of red lake with lead white, calcite and powdered glass and quartz in a matrix rich in oil. In a sample of the dress shadows were detected two different types of red lake since his fluorescence and distribution in the binder is different under optical microscopy with visible and ultraviolet light.

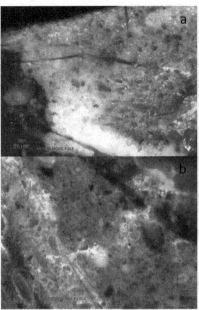

Figure 2. Paint layers into the pinkish red dress of the woman. Cross sections of the chaotic mixtures of red lake and lead white. (a) Polarized optical microscopy, above the union between shadows and lighter colors, 50x, below ultraviolet light (b), two red lake tones, F01, 50x.

The composition of glass particles identified in the Murillo's paintings attribution contains silicon, oxygen, calcium, with small amounts of potassium and aluminum (Figure 3). From this chemical composition, this type of glass was made most probably from silica with wood ash as flux and lime as a stabilizer agent. In comparison, the elemental contents of quartz are only silicon and oxygen.

Spanish historical art treatises as *El Arte de la pintura* of Francisco Pacheco and *El Museo Pictórico y la escala óptica* of Antonio Palomino [4-5], mention the use of grounded glass as a siccative in oil paint and in red lake layers. It seems, that the combination of glass and oil was another of the 'colors' present in the artist's palette. The question is: Why did the artist add glass to a paint that is abundant in lead white? Is very well known that lead pigments are the best siccative materials for the oil technique. Recent literature discusses the function of glass in red lake oil coats as siccative but also as an aggregate capable to create a full-bodied paint and to increase color depth [6-8].

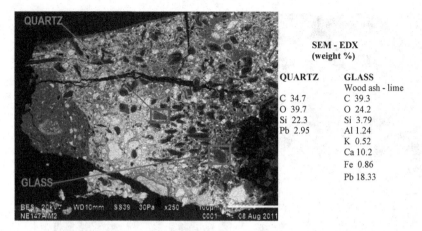

SEM - EDX (weight %)	
QUARTZ	**GLASS**
	Wood ash - lime
C 34.7	C 39.3
O 39.7	O 24.2
Si 22.3	Si 3.79
Pb 2.95	Al 1.24
	K 0.52
	Ca 10.2
	Fe 0.86
	Pb 18.33

Figure 3. SEM image of the colorless fillers in the pink dress layer and the chemical composition of the glass and quartz particles.

EXPERIMENTAL REPRODUCTION AND PAINTING ANALYSIS

In order to understand the real function of the colorless fillers added into red lake layers, we performed a painting experimental reproduction (Figure 4) using three different types of red lakes, made from the most common dyestuff sources in the 17th century: Brazilwood (*Caesalpinia* L.), Madder lake (*Rubia tinctorum* L.) and Cochineal (*Dactylopius coccus* Costa). We tested the commercial lakes of Brazilwood and Madder fabricated by Kremer [9] and also the Mexican brand Tlapanocheztli for the cochineal. The lakes were mixed with four types of transparent glass from both different sources and different manufacture process (Figure 5). Also translucent quartz was tested.

The painting trials were extended in a single coat over a preparatory layer made of lead white bounded with sun-bleached linseed oil. Admixtures of glass into red lake layers were prepared in a proportion of 35% glass and 65% red lake. In all cases, the transparent aggregates increase the thickness of the mixture. Each type of mixture required a different preparation process. Some were achieved by heating the dispersion. Blown glass was the easier material to grind and mix with linseed oil. On the contrary, the glass flakes produce a solution very difficult to blend with the binding medium. Comparatively, quartz coats showed the highest rate of dryness. As the three lakes had the same behavior in the tests, only cochineal was chosen to discuss the influence of glass and silica powder on the thickness and color of an oil film.

The same methods were used for the characterization of original painting samples and laboratory-prepared painting tests: Optical and fluorescence microscopy (we used a Microscope Axiotech Zeiss with halogen lamp for reflected light and HBO 100 mercury lamp for the UV with reflector slide 01 y 05 filter) and SEM was undertaken with a JEOL JSM 6610 LV equipped with a OXFORD INCAx-act for EDS microanalysis. For color measurements UV-VIS Spectroscopy a USB4000 spectrometer of Ocean Optics with an optical fiber of 600 μm diameter and a white color reference for the equipment calibration was used. An halogen light source from Ocean Optics was also used in all the measurements after 25 min of heating.

Figure 4. Laboratory prepared painting tests. From the top to the bottom: Madder Lake, Brazilwood and Cochineal.

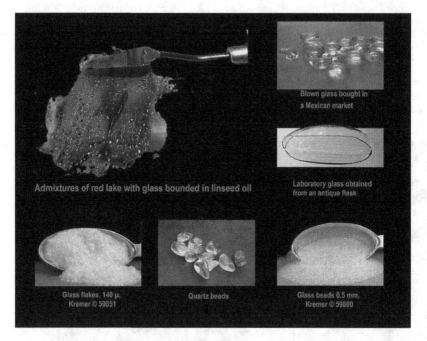

Figure 5. The mixtures for the tests were applied over different grounds. Four types of glass and quartz beads used for the experiment.

DISCUSSION

Size, shape and chemical composition of each type of glass contributes to the qualities of paint layers (Figure 6). Even though the glasses were grounded in the same manner in a mortar and sieved in a measuring range below 52 μm the particle size resultant is inherent to the glass manufacture. The blown glass and the laboratory samples are broken in a more homogeneous size particle around 50 μm. On the contrary, glass flakes and quartz have the largest and the shorter size particles, as a result of their grinding process and their chemical structure. The finest are below the 1 μm but quartz has particles up to 63 μm and flakes up to 70 μm.

Jagged shape and typical laminate form of glass influences the capacity of oil paint to create a film. During the application and drying of the mixtures, the glass particles are oriented according to their form. Glass beads and flakes are precipitated and aligned with the preparation surface. The powdered glass is grouped in conglomerates so when is applied, the thick oil film conserves the print of the brushstroke and it's conserved during the drying process.

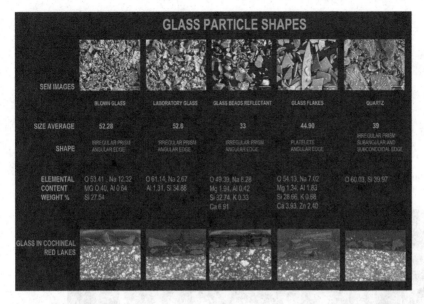

Figure 6. Comparison of size, shape and chemical composition of the powdered glass and quartz particles. The row below shows the material distribution in the Cochineal lake layer.

About the thickness of the paint layers is significant that the cochineal lake oil paint without glass is a very regular film, flat and thin, below 10 μm. In comparison, the glass mixtures multiply the thickness of the oil layer. The thicker layer is obtained with glass flakes and laboratory glass, up to 50 μm. While the thinnest layer correspond to glass beads, around 35 μm. Painting with fillers increases thickness up to four times more than just the blend with red lake and linseed oil (Figure 7). Only the mixture with glass flakes produces a heterogeneous layer. The capacity of oil medium to form a continuous film is difficult because the particle shape. The painting surface is irregular in all cases.

Blown glass, laboratory glass and quartz, are capable to make a good film system because of their size average, jagged form and most homogeneous particle fracture when grounded. These glass types are filling the paint layer and the mixture is a medium that permits to achieve a soft surface and a more homogeneous glaze.

Under UV lighting, we can observe the surface homogeneity of the painted layers. For example, the paint made of red cochineal lake without glass is more transparent than the one with glass added (Figure 8). The filler contributes with the hiding power of the mixture. The more opaque material is the laboratory glass.

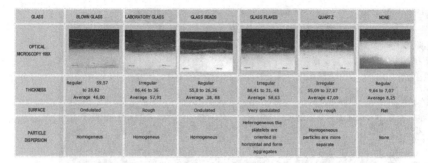

GLASS	BLOWN GLASS	LABORATORY GLASS	GLASS BEADS	GLASS FLAKES	QUARTZ	NONE
OPTICAL MICROSCOPY 100X						
THICKNESS	Regular 59.57 to 28.82 Average 46.00	Irregular 86.46 to 36 Average 57.91	Regular 55.8 to 26.26 Average 38.88	Irregular 88.41 to 31.48 Average 58.63	Irregular 55.09 to 37.87 Average 47.09	Regular 9.64 to 7.07 Average 8.25
SURFACE	Ondulated	Rough	Ondulated	Very ondulated	Very rough	Flat
PARTICLE DISPERSION	Homogeneus	Homogeneus	Homogeneus	Heterogeneous the platelets are oriented in horizontal and form aggregates	Homogeneous particles are more separate	None

Figure 7. Cochineal paint layers with glass and quartz fillers thickness comparison.

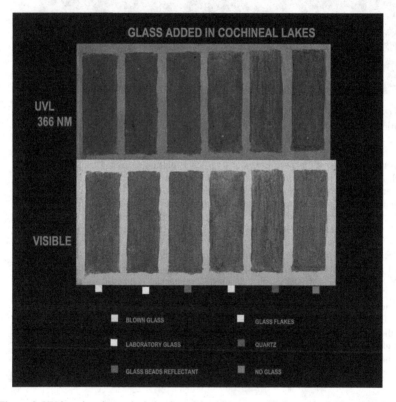

Figure 8. UV imaging shows the behavior of the mixtures applied over the white ground.

Concerning the color measurement of the experimental reproductions and the light dispersion by the paint layers, it is clear that the powdered glass and quartz is changing the optical properties of the paint layer. Figure 9 shows the measurements of light dispersion in the range of 300 to 1000 nm. We can observe in the graph that the color base made with oil red lake layer without any glass addition (BCN) has a blue-green component in the 450-580 nm. This is not present in the paints were glass and quartz were added. Moreover for the layer with quartz (BCC) the red component at 650 nm reduces notoriously. There is not an important difference for the other fillers. The visual total effect of the filler is then to clean the red color of other color components increasing the red components.

The color measurements in the L*a*b* CIE system shows a clear change in the color of the paints (Figures 10 and 11). There is an increment of at least 5 units in the a* (red) and b* (yellow) but a similar decrease in the luminosity (L*) by comparison to the paint without fillers. The redder layer is obtained with glass beads (BCVP), but it also produces a less bright color. A similar behavior occurred for the glass flakes but the most opaque paint is produced with the laboratory glass (BCVL).

Finally, the glass has a direct impact in the color intensity of the red lake paint, it increase the saturation or the purity of color. Because of that, the artists commonly used red lakes as a glaze, a translucent coat that increases the color brightness.

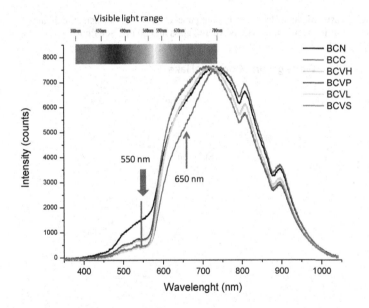

Figure 9. Light dispersion spectrum of the cochineal paint tests applied over a white lead ground. The black line is the mixture without any filler (BCN). Quartz (BCC), glass flakes (BCVH), glass beads (BCVP), laboratory glass (BCVL), blown glass (BCVS).

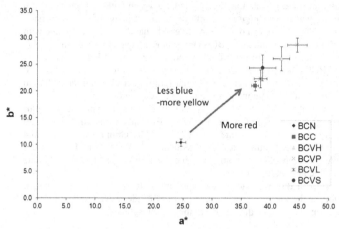

Figure 10. Color measurements of the a* & b* components showing the change and gain of red and yellow color in the mixtures with powdered glass and quartz. See figure 9.

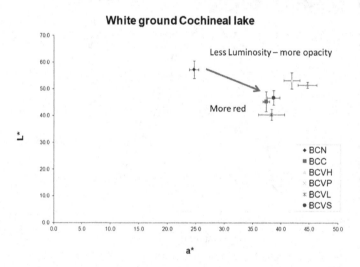

Figure 11. Color measurements of the a* & L* components showing than for all the painting tests with fillers the color is redder and more opaque than the red lake in oil medium. See figure 9.

CONCLUSIONS

In this study we discovered some of the tricks in the handling of paint which aim was to create the most beautiful and pure red color. We discuss around the reasons of adding powdered glass and quartz into the oil films as a part of the painting making.

Grounded glass and grounded quartz, being transparent raw materials, both give to oil layers different physico-chemical properties: first, they are functioning as aggregates that increase the thickness of the oil film and improve the cover properties of the color by means of their capacity to produce a more opaque layer. Second, even if they are good siccative for oil, they are usually associated with lead white so its presence is more related with an aesthetic function in oil paint. Third, they are capable to enhance the brightness of red color in oil red lake paint –virtually cleaning the color of other components in the visible range- because of the higher light reflectance in layers with homogeneous glass particle distribution.

In the Murillo's attributed paintings, more than the siccative function, quartz and glass, both were intentionally added to intensify the brightness and to modify the color saturation. These fillers make redder the lake glazes and were employed to create a full-bodied paint.

ACKNOWLEDGMENTS

The authors would like to thank Said Rico art student from the Escuela Nacional de Artes Plásticas, Universidad Nacional Autónoma de México. He prepared the materials in the laboratory and applied the paint trials. This research has been carried out with support of UNAM PAPIIT IN403210 ANDREAH project as well as the CONACYT Mexico grant 131944 MOVIL II. This study has been performed in the frame of the Non Destructive Study of the Mexican Cultural Heritage - ANDREAH research network: www.fisica.unam.mx/andreah.

REFERENCES

1. F. S. Gutiérrez, Impresiones de viaje, Juan Panadero, Guadalajara, VII, n. 456, 21 de diciembre de 1876, Ref. A. Camacho, *Los papeles del artista*, El Colegio de Jalisco, Zapopan, 2010, 41.
2. J. G. Zuno, L. R. Zaragoza, *Los enigmas del Museo de Guadalajara*, Guía del Museo de Guadalajara, Ediciones Centro Bohemio, México, 1950.
3. E. Arroyo, E. Hernández, *Informe técnico de dos cuadros sobre la vida de San Francisco atribuidos a Bartolomé Esteban Murillo pertenecientes al acervo del Museo Regional de Guadalajara, México*, Laboratorio de Diagnóstico de Obras de Arte, Instituto de Investigaciones Estéticas, UNAM, UNAM, 2011. (unpublished)
4. F. Pacheco, *El Arte de la Pintura*, [First edition 1649], Cátedra, Madrid, 1990, 484.
5. A. Palomino de Castro y Velasco, *El museo pictórico y escala óptica* [First edition 1715-1724], Imprenta de Sacha, Madrid, 1796, 56.
6. K. Lutzenberger, H. Stege, C. Tileschi, *Journal of Cultural Heritage* 11 (2010) 365-357.
7. M. Spring, Raphael's materials: Some new discoveries and their context within early sixteenth-century painting, in: *Raphael's Painting Technique: Working Practices before Rome*, A. Roy and M. Spring (Eds.), Proceedings of the Eu-ARTECH Workshop,

Quaderni di Kermes, London, November 2004, Nardini Editore, Florence, 2007, 75-84.

8. M. Spring, Perugino's painting materials: analysis and context within sixteenth century easel paintings, in: *The painting technique of Pietro Vannucci, called il Perugino*, B.G. Brunetti, C. Seccaroni, A. Sgamellotti (Eds.), Proceedings of the LabS-TECH Workshop, Quaderni di Kermes, Perugia, April 2003, Nardini Editore, Florence, 2004, 21-28.

9. Kremer code number 36160 Extract of Brazilwood and 37202 Madder Lake genuine. http://kremer-pigmente.de

Mater. Res. Soc. Symp. Proc. Vol. 1374 © 2012 Materials Research Society
DOI: 10.1557/opl.2012.1379

Xavier Guerrero, De México a Chile (From Mexico to Chile), Some Remarks About the Use of Portland Cement in Mexican Muralism

Sandra Zetina Ocaña[1], Manuel Espinosa Pesqueira[2], Nora Ariadna Pérez Castellanos[3], Renato Robert Pappereti[4]

[1] Laboratorio de Diagnóstico de Obras de Arte, Instituto de Investigaciones Estéticas, Universidad Nacional Autónoma de México, UNAM. Circuito Mario de la Cueva s/n, Ciudad Universitaria, México DF 04510, Mexico. e-mail: sandra.zetina@gmail.com

[2] Instituto Nacional de Investigaciones Nucleares ININ, Mexico

[3] Instituto de Investigaciones en Materiales, Universidad Nacional Autónoma de México, UNAM. Mexico.

[4] Centro Nacional de Conservación y Registro del Patrimonio Artístico Mueble, Instituto Nacional de Bellas Artes, INBA, Mexico.

ABSTRACT

Xavier Guerrero (1896-1974) had an important role in the so-called Mexican Mural Renaissance, as a technical leader in the murals painted by Roberto Montenegro and Diego Rivera in the early 1920's. Jean Charlot, Diego Rivera and David Alfaro Siqueiros considered him as a sophisticated fresco craftsman, whose knowledge came from a popular mural painters guild.

In 1941 the Mexican Government donated a School to Chillan, a Chilean town almost destroyed by a strong earthquake. David Alfaro Siqueiros and Xavier Guerrero were commissioned to paint murals on the *Mexico School*. Between 1941-1942 Guerrero decorated several walls and the staircase ceiling, the mural program is called *De México a Chile (From Mexico to Chile)*. In 2010 another earthquake destroyed part of the ceiling. This study is part of the diagnosis project of *De México a Chile*, and consists in the characterization of the mortar and painting layers with optical microscopy, X-ray diffraction (XRD), scanning electron microscopy (SEM) and thermal analysis (TGA), while textural properties of the mortars were studied with nitrogen adsorption-desorption techniques.

Analytical results show a stratigraphic sequence composed of several layers of Portland and lime combinations, and also an interesting painting technique that possibly involves the Portland cement setting process, with the development of specular gypsum.

INTRODUCTION

Two earthquakes marked Chillan recent history; both are related to Xavier Guerrero mural *De México a Chile* at Mexico primary School. The first earthquake prompted the construction of the primary as a gift of Mexican people to Chilean people, with the mural decoration and the second badly damaged the murals.

The first earthquake strike the city on January 24th, 1939; it had an intensity of 8.3 in the Richter scale, it was one of the strongest ever felt in the region, it left a dead toll of 30,000 people, and almost destroyed the city. As a contribution, the Mexican government donated the Mexico School, a primary that emphasized the importance of education in the political program of the left wing leaders from both countries: Lázaro Cárdenas and Pedro Aguirre Cerda. Architect Eduardo Carrasco won the public design contest; he placed the first stone in April 1940. [1] A few months later, the Mexican muralist David Alfaro Siqueiros, involved in an assault to Leon Trotsky, was commissioned by Mexican president Manuel Ávila Camacho to paint a mural at the Mexico School in Chillan, a safe way to get out of the international problem of liberating the communist artist. Siqueiros proposed a mural to be painted at the library: *Death to the invader*, an allegory of Mexico and Chilean independence and revolution. While Siqueiros was painting *Dead to the invader*, Xavier Guerrero, also a communist artist, was invited in 1941 to Chillan to contribute with the mural program. Guerrero chose the transit spaces to paint his murals: the main hall and the staircase. Also, some Chilean artists depicted historical leader's portraits throughout the corridor, which were destroyed some time after [1].

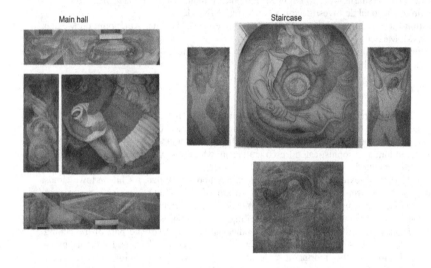

Figure 1. Xavier Guerrero, *De México a Chile (From Mexico to Chile)*, México School, Chillán, Chile, 1941-1942, left, murals at the main hall walls, and in the center murals in the ceiling *(Forces of nature* and *A Mexican Mother helping a child)*. Right, panels distributed over the staircase, at the center the plafond mural *La historia y el hombre (Man and History)* and below, *Gobernar es educar (To rule is to educate)*. The titles were mentioned by a Maria Izquierdo's contemporary critique [2].

Xavier Guerrero murals had localized conservation treatments between 1942 and 2009. In 2009 an international collaboration between Mexican and Chilean institutions, Centro Nacional de Conservación y Registro del Patrimonio Artístico Mueble from Mexico and Centro Nacional de Conservación y Restauración de Chile from Chile, designed a restoration project for Siqueiros and Guerrero murals. The restoration was finished by October 2009. Shortly after, in February 27th 2010 a 8.8 Richter degree earthquake hit Chillán again, causing loses of 20% of the Xavier Guerrero mural at the staircase plafond.

Xavier Guerrero had a particular role in Mexican muralism, he was not a prolific mural painter, but participated as the fresco technical master, he was considered the expert by all the Mexican muralists: Diego Rivera, David Alfaro Siqueiros and Jean Charlot [3]. At the murals in Chillan he used two different techniques one for the walls of the hall and another for the plafonds that cover the stairway. None of them seems to be traditional buon fresco, so here, we will discuss first the composition and structure of the mural, and then a brief consideration of fresco as a technique in modern Mexican muralism.

MURAL TECHNICAL DIFFERENCES

De México a Chile is a mural program of about 128 m2; it is divided in two spaces of the same corridor, the hall at the entrance of the school and the stairway plafond. The main hall walls depict people suffering under the uncontrollable elements of nature, and science, solidarity or education as a way to cope with them. As a closure of the program, the plafonds at the staircase present allegories of offering and history.

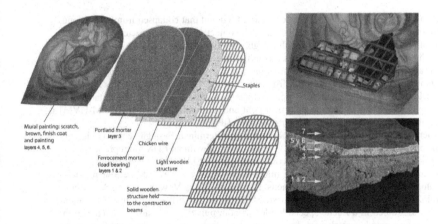

Figure 2. Left, structure of the plafond in *La Historia y el Hombre (Man and History)*; upper right, the collapse of the lower part of the plafond; below, the fragment of plafond studied, some layers are evident macroscopically: (1), (2) ferrocement mortar, (3) Portland cement coarse coat, (4) brown coat, (5) and (6) skim coat, (7) painting layer.

Guerrero chose a different technical approach for each of the spaces, possibly due to the building technique of the support: the open wide hall was built with bricks and Portland cement, and the stairway ceiling was a floating structure. The section that suffered more damages during the 2010 earthquake was the staircase plafond.

The stairway mural, named *La historia y el hombre (Man and History)* depicts a brawny woman, seated over a rainbow, she grabs flames and holds a rolled document on the right hand. The center of the composition is a gigantic floating hand that holds a compass with the inscription "The most precious treasure is the man". In this panel Guerrero used other technical strategies besides painting, he manipulated the mortar in relief, for example in the compass perimeter, or he used incisions to draw up the boundaries of the figures; in some areas, he used incisions instead of lines, engraving the design over a red painted surface, therefore obtaining a negative image that imitates an etching on copper. In the presentation of the Chillán murals, written in 1943 for the inauguration, the Chilean writer Ernesto Eslava mentions that Xavier Guerrero had some technical difficulties with the plafond, and he recalls that Guerrero had to invent a new technique [4].

The plafond was made of a kind of ferrocement: chicken wire joined to a light wooden structure with staples. The mesh of wire was covered with a thick mortar layer. No steel rods or rebars are visible; instead, the mixture is attached to a wooden structure that constitutes the plafond (see figure 2). Then it was plastered with scratch, brown and skim coat, each of them are constituted by mixes of Portland cement, sand and lime in different proportions. Part of this structure collapsed.

METHODOLOGY

Two samples were studied, one taken from a fragment that collapsed from the plafond, and another taken directly from the hall walls. The objective was to characterize the mortar composition; the plafond sample was thoroughly studied to have relevant information to design the conservation and anastylosis process of the collapsed area. The fragment from the mural was of approximately 15 cm long and 7 cm wide; the sample included the ferrocement mortar. The sample taken from the walls of the main hall only has the mural process layers. Half of the samples were prepared in cross sections to understand their stratigraphy and microstructure in polarized light and ultraviolet microscopy (PLM-UVM) at the Laboratorio de Diagnóstico de Obras de Arte, IIE-UNAM.

Microstructure and elemental composition was determined through scanning electron microscopy with a tungsten filament SEM 6610 LV fitted with an EDX microprobe Oxford with a Si Drift detector; experimental conditions were 20 kV, 70μA, low vacuum mode, low vacuum pressure 20 Pa inside the microscope chamber. X ray diffraction provided mineralogical composition (XRD Siemens D-5000 at 35 KeV with a 25 mA current). Several XRD techniques were used: to study the pictorial surface a sample with a flat surface was studied in a Lucite special holder without pulverizing it; also different mortar strata were separated: scratch coat, brown coat, and skim coat (*rinzaffato, arricciato* and *intonaco* as they are called in traditional buon fresco treatises) each of them was examined through the powder conventional method. The analysis was performed in a range from 2.5° to 70° 2θ degree, with intervals of 0.019°, and step time of 1 sec. The crystalline phases were identified with the JCPD-ICDD Version 1.2 cards. SEM and XRD were done at Laboratorio de Microscopía Electrónica y DRX, Instituto Nacional de Investigaciones Nucleares.

Thermogravimetric (TGA) measurements were carried out on a TGA Q5000 from TA Instruments, the measurements were obtained at a heating rate of 15°C min-1. A flow rate of 25 mL min-1 of nitrogen was used for all experiments. Nitrogen adsorption measurements at 77 K were performed on an BELSORP-miniII volumetric adsorption analyzer. Before the measurements, the samples were outgassed at 308 K for 20 h. TGA and specific surface area (BET) were done at Instituto de Investigaciones en Materiales, UNAM.

RESULTS

Figure 3 shows the cross sections of the studied samples. The data corresponding to these samples is presented in Table I.

In optical microscopy the comparison of the stratigraphic sequence of the plafond sample and the wall sample show some similarities, an analogous variation in texture and width in the brown coat and the skim coat. In both samples the *arrciato* or brown coat (layer 4) is very thick, and combines lime with Portland cement and natural volcanic sand (subangular particles); the *intonaco* or skim coat is extremely thin in relation with the *arriciato*, and in both cases is palpable the change in composition of the binder and aggregates (an increase of lime and whiter, smaller sands). Table I synthesizes the function of each layer, the size and composition of its aggregates and matrix obtained through XRD, SEM and optical microscopy.

Is evident that the plafond sample has more layers due to the ferrocement mortar, it has seven layers, and the hall mural only four. The first two layers in the plafond mural constitute the ferrocement, probably done by masons following the architect directions; this mortar has a better distribution of particle size. It is possible that Guerrero guided the preparation of the plafond surface, he may suggested the application of the coarse coat, a mortar rich in Portland cement, which has no structural function, but may have been used as an adhering layer of the subsequent coats. The differences in the proportion of binders (lime, Portland cement and gypsum) are intended to achieve particular properties in the resultant mortar.

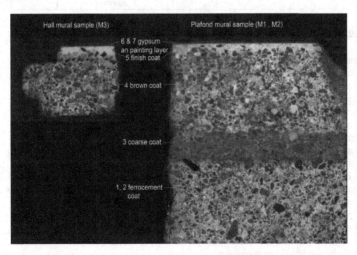

Figure 3. shows the cross sections of the studied samples.

Table I. Synthesis and comparison of the hall mural and plafond stratigraphic sequence; (microphotograph of both samples, 1x).

Number and function	Hall mural sample (M3)	Width	Plafond mural sample (M1, M2)	Width
(7) Painting layer	Chrome green and synthetic ultramarine	5-15 μm	Chrome green	0-3 μm
(6) Crystallization layer (only visible in SEM)	Calcium carbonate layer with some silica	3-5μm	Specular gypsum	6.5 μm
(5) Skim coat *intonaco*	Lime with Portland cement. Siliceous aggregates (10-80 μm)	500-1000 μm	Lime with less than 1% of Portland cement, some gypsum. Calcareous round aggregates (10-80 μm) and some igneous fine sand	850-1000 μm
(4) Brown coat *arricciato*	Lime with Portland cement, a low proportion of gypsum. Igneous sand and calcareous aggregates (40-600 μm)	4 mm	Lime enriched with Portland cement, a low proportion of gypsum. Igneous sand and calcareous aggregates (40-600 μm)	6-7 mm
(3) Coarse coat *rinzaffato*			Portland cement, Igneous sand (400-1500 μm)	3.5 mm
(1) (2) Ferrocement mortar			Portland cement and sand in low proportion. Igneous sand (40-1500 μm) In optical microscopy this layer is composed of two applications of an identical mortar	45-70 mm

Crystalline phases

A fragment of the plafond mural sample was separated in layers for the mineralogical characterization through XRD (Figure 4). The crystalline phases found in each stratum are presented in Table II. The ferrocement coat has feldespars, anorthite in high proportion due to volcanic aggregates, also calcite, quartz and iron oxides (present in aggregates and also in the Portland cement). The next layer, the coarse coat, in addition to the calcite and quartz phases, contains calcium and sodium aluminosilicates that indicate a higher proportion of Portland cement; the sand composition is mainly albite and anorthite. In the brown coat there is a strong signal of calcium aluminosilicate phase indicated a high proportion of Portland cement, even though it has calcite and a very small amount of gypsum (less than 3% in weight, only detected by SEM). The sands present in this layer are similar to those in the coarse coat, composed of volcanic sand, identified by the albite and anorthite phases. The finish coat, the pictorial layer and the gypsum layer, analyzed altogether by the conventional powder method contain calcite, present quartz and gypsum as the main crystalline phases, there instead of igneous sand Guerrero added marble powder or calcareous aggregates. The phases present at the pictorial layer indicate that it was achieved with an iron and chrome oxide green pigment.

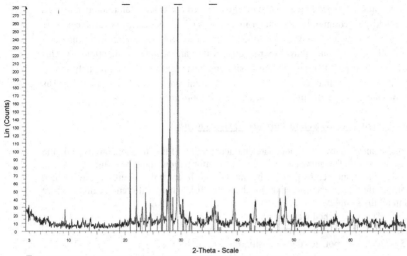

Figure 4. shows the XRD pattern of one of the layers of the mural sample.

Table II. XRD of the plafond samples (M1, M2), the numbers represent a relative proportion of each phase, 1 indicating the highest percentage and 4 the lowest. Also the spectra of brown coat (layer 4) is presented.

Crystalline phase / Layer	Calcite CaCO$_3$	Albite Na (AlSi$_3$O$_8$)	Anorthite Ca Al$_2$ Si$_2$O$_8$	Calcium aluminosilicate Ca$_{0.88}$ Al$_{1.77}$ Si$_{2.23}$O$_8$	Calcium and sodium aluminosilicate (Na$_{0.45}$Ca$_{0.55}$) (Al$_{1.55}$Si$_{2.45}$O$_8$)	Quartz SiO$_2$	Gypsum CaSO$_4$(H$_2$O)2	Iron oxide Fe$_2$O$_3$	Iron and chrome oxide Cr$_{1.3}$Fe$_{0.7}$O$_3$
Ferrocement coat, layers 1 & 2	3	—	1	—	—	2	—	4	—
Coarse coat layer 3	2	4	5	—	1	3	—	—	—
Brown coat layer 4	1	2	—	4	—	3	—	—	—
Skim coat layers 5, 6 & 7 (powdered)	1	—	—	—	—	3	2	—	4
Pictorial layer, layers 6 & 7	2	—	—	—	—	3	1	—	4

The sands may have been local, the anorthite detected is a calcium aluminum silicate, a plagioclase feldspar, commonly present in metamorphic or igneous rocks, they could be from the region of Talcahuano or near Chillan, in Bio Bío province where three important deposits of natural sands are, gravel extraction pits that are actually in function [5]. This kind of alkaline feldspars that come from felsics and basic igneous rocks, probably basalt, granit, riolites or lava that conformed the continental cortex, and that by digenesis (chemical dissolution), weathering and transport turned into sands.

Thermogravimetric Analysis and Porosity Measurements

Thermal analysis allowed an adequate characterization of the mortars employed on the mural (figure 5), the TG curves gave significant information about the similarities and differences of the manufacture process by showing different temperature regions, which correspond to thermal decomposition mechanisms [6]. Three different regions where observed in all the samples:
- 35-150°C   Absorbed water
- 150-550°C   Cement dehydration
- 550-730°C   Carbonates decomposition

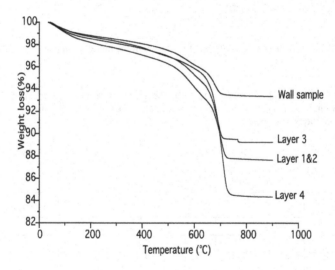

Figure 5. TG plot of the ceiling's layers and the wall sample.

For the different layers of the plafond ceiling sample we have an average of 1.2% of absorbed water, while the weight loss from the evaporation of the water structurally bound to the hydraulic components was different for each layer: for layers 1 and 2 the weight loss was of 4.5%, for layer 3 it was of 3.5%, finally the layer 4 lost 4.2% of its weight. The rate of evaporation was similar for all the layers indicating that the materials employed were the same (see figure 4), since the behavior of the curves in this zone is affected by the fillers and their capacity of trapping water in their structure [6].

The third region which corresponds to the amount of carbonates was the one that had major differences between the layers, this showed how the artist changed the proportion binder to filler for obtaining different textures and hardness required for working on the mural (layers 3 and 4).These results are in agreement with the XRD analysis since the percentage of carbon dioxide is higher for the fourth layer. The percentage of CO_2 for layers 1 and 2 is of 7.8%, for layer 3 is 6.8% and for the last layer is 10.2% . The TG curve in figure 3 has different slopes in this region, which can be attributed to the multiplicity of carbonate species due to the nature of the fillers [7,8], we can see how this behavior is similar on layers 1,2 and 3 while in the last layer this contribution is very small due to the decrease of the fillers as shown in XRD. Since the decomposition temperature of the calcite in the samples is around 650°C it is possible that the calcite present is a polycrystalline material [9].

The sample taken from the wall follows the general trend but has different rates of dehydration which is attributed to the nature of the filler as shown in the SEM images. The weight loss due to absorbed water is of 1.0%, 2.9% is due to the dehydration of the mortar and finally this sample lost 3.5% for the carbonates decomposition. The fact that the decomposition temperature is the same for the calcite indicates that this material probably comes from the same source, while the fillers in the ceiling sample have a different provenance than the ones at the wall sample; therefore indicating different techniques and temporalities (the hall murals were made first) in the murals.

According to Rizzo et al. [10] it is possible to determine the hydraulic character of the mortar by obtaining the ratio of CO_2 and absorbed water, in figure 6 the series in black corresponds to the CO_2/ H_2O_{abs} axis, the lower the value is on this correlation the more hydraulic is the mortar. The samples that belong to the structural part of the mural such as the wall sample and the first three layers of the ceiling sample show a similar trend varying the hydraulic character. The layer 4 from the ceiling sample since it is mainly composed of lime it is an air mortar and it is shown at the top of the graph.

Physical adsorption is one of many experimental methods available for the characterization of porous materials, since gas adsorption allows assessing a wide range of micro- and mesopores.

The samples of the mural show a type II adsorption isotherm and a type H4 hysteresis loop according to the IUPAC classification [11] with the hysteresis loop closing at about p/po= 0.35, this type of hysteresis is due to narrow slits pores and also micropores. Table III presents the values for the surface area calculated by the BET method, the results agree with those presented for Portland cement in reference 11 [12].

As it was expected the values obtained for the wall sample are higher than the average value obtained for the ceiling sample, this is in agreement with the bigger particle size of the filler showed in the microscopic analysis and the calcite content showed in the TG (and XRD) results hence increasing the surface area and pore diameter.

Again through this analysis it is showed the differences in the materials employed for the ceiling and the wall.

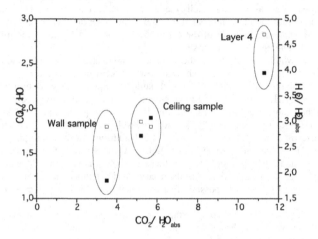

Figure 6. Representation of hydraulic character in the two mortar samples. White squares correspond to the H2O/H2Oabs axis , black squares correspond to CO2/H2O axis.

Table III. BET$_{N2}$ specific surface area and pore diameter calculated by the BJH method, for both the ceiling and the wall sample.

Sample	Specific Surface Area $_{BET}$ (m²/g)
Layers 1&2	3.5
Layer 3	3.7
Layer 4	6.3
Wall sample	8.1

Pictorial layer

The most interesting difference between the ceiling and the wall mural technique are the skim coat and the painting layer composition and structure. In the ceiling the presence of a crystallization layer between the skim coat and the pictorial layer was noticed. It is a thin layer of oriented specular gypsum, which contrasts with the presence of a calcium carbonate layer on the wall sample (see table IV).

Table IV. Hall mural sample (M3) and ceiling plafond mural sample (M2) SEM-EDS comparison of elemental composition.

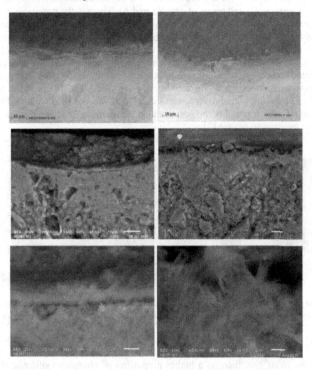

Element	Ceiling finish coat layer 5 wt%	Hall finish coat layer 5 wt%	Ceiling gypsum layer 6 wt%	Hall calcium carbonate layer 6 wt%	Ceiling painting layer wt%	Hall painting layer 7 wt%
C	21.4	10.7	22.4	24.5	28.4	21.9
O	49.9	55.6	55.4	52.8	42.4	52.5
Na	-	-	-	1.42	0.65	1.61
Mg	0.34	-	-	-	0.38	-
Al	0.41	2.08	0.25	1.08	1.04	1.50
Si	3.28	7.2	0.64	2.76	1.63	2.91
S	1.44	0.22	**5.24**	**0.96**	2.13	1.32
K	-	0.28	-	0.32	0.20	0.46
Ca	22.9	24.0	**16.0**	**16.1**	16.2	14.8
Cr	-	-	-	-	2.77	3.00
Fe	-	-	0.06	-	4.45	-

In XRD a gypsum phase was identified along with quartz, calcium carbonate and chrome iron oxide. In the surface it seems like a whitewash of very fine particles of gypsum, with specular morphology. This is not an alteration product because the painting as well as the skim coat are in perfect conditions. The chemical elemental analysis (SEM-EDS) of the pictorial layer in the ceiling sample showed a heterogeneous composition: a high proportion of Ca, with S, Si, and traces of Mg, Na and Al. There is the possibility that Xavier Guerrero used a gypsum finishing whitewash, but it is also probable that this perfectly oriented gypsum particles are part of the process of painting in "fresco" over mortars that have some cement proportion. In both of the samples the Si is always present, due to some silicieos aggregates, but also in the intonaco or skim coat matrix. It is evident that the painter wanted to achieve different effects at each region (see figure 7). The art critic Antonio Rodriguez mentioned that Guerrero knew a very special traditional painting technique that involved gypsum, a kind of "gypsum fresco" that was polished and dried by ironing. [13] Also Jean Charlot, possibly the first muralist that painted with fresco in early 1920's, mentions this gypsum technique [3].

Figure 7. Technical differences evident in the kind of finishing, left, wall murals, center relief effects and right incisions instead of transparencies in the ceiling mural *Man and History*.

In the wall sample skim coat there is a higher proportion of aluminum silicates, the proportion of Si and Al significantly increases, along with calcium carbonate. Some sulfates were present but over the painting layer, more as an alteration process (see table IV).

CONCLUSIONS

The technique of Xavier Guerrero mural *De México a Chile* presents two variations of the traditional fresco technique, using Portland cement and possibly gypsum. The hall wall murals are more porous, and have a strong proportion of cement even in the intonaco or skim coat layer, but the crystallization process produced calcium carbonate that holds the pigments together. These murals were done first, and over brick walls, but in the second section of the program, at the staircase Guerrero decided to modify his technique, he innovated the technique possibly also through the application of a gypsum layer. SEM,

TGA, BET and XRD permitted to define each layer composition, morphology and porosity. Also it is evident that Guerrero used local sands and there is the possibility of some collaboration with the architect Eduardo Carrasco because at the plafond he used very similar materials as the construction ones.

ACKNOWLEDGMENTS

The authors would like to acknowledge the support and collaboration from Gabriela Gil, director of CNCROPAM-INBA, also the studies and research applied to the 2010 conservation project given by the specialists from Instituto de Ingeniería, UNAM. We also thank the support in logistic from Mexican Embassy in Chile to the research. The contribution from David Oviedo, coordinator of the conservation Project 2008-2009 in Mexico School and Elena Acosta, both restorers from CENCROPAM. To Paula Riquelme, Daniela Gonzáles, Ernesto Meza, Claudio Cofré, Chilean undergraduate art students that participated in the conservation Project, and also to conservators Mónica Pérez and Eduardo Walden from Centro Nacional de Conservación y Restauración de Chile.

REFERENCES

1. F. Torres P., R. Vera M. L. Arias E., *América es la casa, Arte mural y espacio público en Chillán, Chile,* Consejo Nacional de la Cultura y las Artes, Municipalidad de Chillan, Asociación Chilena de Seguridad Chillán- Copelec, 2011, 31-32.
2. M. Izquierdo, Arte: La pintura mural de Xavier Guerrero en Chillán. *Hoy*, Mexico City, 282 July (1942) 46-49.
3. J. Charlot, La Época de Xavier Guerrero, *Época: México en la Cultura,* México, February 27, No. 1/196, (1972) 5-7.
4. Ernesto Eslava, Ofrenda de México, in *Pintura Mural*, Escuela México, Chile, 1943.
5. M.C. López O., A. Gajardo C., R. Carrasco O., J.L. Mendoza F., Yacimientos de rocas y minerales industriales de la VIII Región del Bio Bio in: *Carta Geológica de Chile, Serie Recursos Minerales y Energéticos*, no. 16, 1:500.000, Servicio Nacional de Geología y Minería de Chile / Subdirección Nacional de Geología, Santiago, 2003, p.8.
6. D.A. Brosnan, J. P. Sanders; S.A. Hart, *Journal of Thermal Analysis and Calorimetry* 106 (2011) 109-115.
7. R. M. H. Lawrence, T. J. Mays, P. Walker, D. D'Ayala, *Thermochimica Acta.* 444 (2006) 179-189.
8. A. Duran, M.D. Robador, M.C.J. de Haro, V. Ramirez-Valle, *Journal of Thermal Analysis* 92 (2008) 353-359.
9. A. Duran, L. A. Perez-Maqueda, J. Poyato, J. L. Perez-Rodriguez, *Journal of Thermal Analysis and Calorimetry* 99 (2010) 803-809.
10. G. Rizzo, B. Megna, *Journal of Thermal Analysis and Calorimetry* 92 (2008) 173-178.
11. M. Thommes, in *Introduction to Zeolite Science and Practice*, J. Cejka et al. eds, Elsevier Science Publishers, London, 2007.
12. I. Odler, *Cement and Concrete Research* 33 (2003) 2049-2056.
13. A. Rodríguez, Xavier Guerrero, Artista del Pueblo, *El Nacional*, Mexico City, 1949.

Archaeological Science

Mater. Res. Soc. Symp. Proc. Vol. 1374 © 2012 Materials Research Society
DOI: 10.1557/opl.2012.1380

Foreing Produced Shell Objects in the Templo Mayor of Tenochtitlan

Adrián Velázquez Castro[1], Belem Zúñiga Arellano[2] and José María García Guerrero[2]
[1] Museo del Templo Mayor, Instituto Nacional de Antropologia e Historia, INAH.
Seminario No. 8, Centro Histórico, Mexico DF 06060, Mexico.
e-mail: adrianveca62@gmail.com
[2] Proyecto Templo Mayor, Instituto Nacional de Antropologia e Historia, INAH. Seminario
No. 8, Centro Histórico, Mexico DF 06060, Mexico.

ABSTRACT

The analyses of work traces in the shell objects found in the offerings of the Great
Temple of Tenochtitlan, by scanning electron microscopy (SEM), has allowed to find an
important group of objects made locally in Tenochtitlan. These shell pieces have been
found in the constructive stages IVb to VII (1469-1520). Recently, another groups of
objects have been found that present different work traces and that seems to be foreign
productions. In this paper this new data will be presented and it will be discuss the possible
origin of the objects.

INTRODUCTION

Until only a short time ago, scholars of Mesoamerican cultures were of the widespread
opinion that the Mexicas played a relatively small role in the production of their material
culture. Therefore, for example, López Luján has stated that the most abundant pieces in
the offerings found in the Templo Mayor of Tenochtitlan were of foreign origin, probably
acquired through tribute, trade, gift-giving, or looting [1]. However, recent research has
shown that at least a part of the goods interred in offerings in Tenochtitlan were actually
locally produced, some possibly made in the ruler's palace, and intended for state religious
practices [2-4].

Part of the results of this research has also been the detection of groups of pieces that
predate Tenochca imperial production, which seems to have been consolidated starting with
the rule of Axayacatl (1469–1481) [5]. Also identified are elements with features that
clearly indicate foreign origin, which can occasionally be traced. The purpose of the present
paper is to present two groups of shell objects that may be regarded as foreign. In one case
the possible origin of the pieces may be proposed, but in the other, unfortunately, it remains
uncertain.

STUDYING THE STYLE OF SHELL OBJECTS

Since its origin one of the basic concerns of archaeology has been determining the
affiliation and dating of material remains left by past societies. Traditionally both issues
have been dealt with through the identification of diagnostic traits, whose origin could be
traced to a site, a geographic region, a culture, or a period. Underlying this approach is the
assumption that different cultures have particular ways and characteristics of producing
objects, in other words, a specific "style".

Definitions of styles differ depending on the different theoretical trends underlying
them, although in general it may be said that style alludes to the systematic and regulated

choices of known alternatives that are standardized in a recurrent way of presenting forms and processes [6]. Its characteristics can include having a restricted repertoire of shapes that bear a relationship within the corpus or to the whole [7], which may be identified as the sole product of individuals or possessing a temporal coherence in space [7]. Although currents such as the New Archaeology does not grant them any more than secondary importance in the social dynamic, post-processual archaeologists assign it an active role in the communication of messages, regulating forms of behavior, and creating cohesion in social groups, as well as differentiating them from others [8-11].

The problems of attributing affiliation and temporality to objects based on stylistic features have been amply raised by neutron activation analysis practiced on the pastes of a number of ceramic pieces from the Templo Mayor of Tenochtitlan. It was possible to determine, for example, that an anthropomorphic urn from the House of the Eagles, which was originally thought to be of the type known as Tohil-Plumbate - characteristic of the Early Postclassic and originally from the Pacific coast, between the Mexico - Guatemala border - was none other than a local imitation, because its clay sources were in the Basin of Mexico [12]. A similar discovery was made with two urns that were thought to be of the type known as Fine Orange, originally from Central Veracruz, whose raw materials displayed a close relationship with Matlatzinca polychrome ware from the Valley of Toluca and with Mixteca-Puebla vessels found at Taxqueña, in the southern part of modern-day Mexico City, again leading to the conclusion that the clay source must have been from a location west of the Basin of Mexico [13].

In this context, studies of technology have emerged as an option that can enrich our understanding. Technology may be defined as a group of social and material elements with which man modifies his surroundings to satisfy a variety of needs; its importance for the comprehension of human groups is clear, because it shows humankind's relationship with nature throughout his history. Technology includes both tools and products made with them, such as knowledge, behavior, attitudes, and meanings, which are shared by groups of individuals—with time by society as a whole—being transmitted from one generation to another [14-16]. In the field of technology, technical processes—also known as operational chains—are included, which refer to the series of steps followed since the material is intact until the object is completely finished [14, 16-18]. In each one of the phases of these sequences of activities, the producers have to make decisions faced with a spectrum of variable possibilities, restricted by environmental, historical, social, and cultural factors [14-16]. There are no external limitations to a human group sufficiently powerful to be the only causal factors of all decision-making in the operational chains [19]; therefore, it has been suggested that technology is a social construction that reflects aspects ranging from organizational principles to ideological considerations (political and symbolic, for example) [15].

Based on ethno-archaeological studies, it is clear that decisions that are made in operational chains tend to be highly specific and consistent, dictated to a large extent by custom [20]. Similarly, it has been possible to verify how technological limits coincide with those of communities. As a result, the concept of "technological style" has been proposed as the sum of choices a human group makes, which forms the knowledge of a manufacturing tradition [21]. One of the advantages of technological studies in relationship to style is its stability in time, since changes imply modifications in manufacturing processes, which traditional societies are generally reticent to employ [21].

Since 2000 the project "Técnicas de manufactura de los objetos de concha del México prehispánico" (TMOCMP, "Manufacturing Techniques of Shell Objects in pre-Hispanic Mexico") has been conducting research at the Templo Mayor Museum. One of its principal objectives is awareness of the different technological styles that developed at different sites, regions, and periods in the pre-Columbian history of Mexico. For this purpose, experimental archaeology has become a paradigm, given the frequent lack of so-called direct indicators of production for many of the collections of shell objects, which come from ritual contexts. Therefore, modifications (abrasion, cuts, perforations, openwork, incisions, and surface finishing) made in antiquity to manufacture objects are replicated in modern shells of the same species as those from pre-Hispanic contexts. The original materials and tools are gleaned from different sources of information (archaeological discoveries, historical sources, and the work of other researchers) and are assumed to be what was used in the past. To go beyond pure conjecture and to propose with a greater degree of probability the procedures and utensils used, experimental manufacturing traces are characterized and compared with the features observable in archaeological pieces. This is carried out at three levels: macroscopic analysis (examination with the naked eye), stereoscopic microscopy at low magnification (10x, 30x and 63x) and the scanning electronic microscope (SEM) (100x, 300x, 600x and 1000x). The latter is the technique that has produced the best results, as it is ideal for the study of superficial characteristics of materials. For SEM analysis a beam of electrons is projected on the sample to be analyzed, which produces a series of signals that are captured by special detectors, some of which can generate high resolution digital images. In the case of conventional equipment, the sample chambers must be in high vacuum conditions and the samples must be conductors of electricity; in more recent models it is possible to work in low vacuum conditions and even at environmental pressure, which makes it possible to analyze moist, organic samples. Initially the study of shell objects with SEM was conducted by moving the pieces to the lab. Later the decision was made to obtain replicas of manufacturing traces in polymers softened with acetone—a technique used in metallography—which are covered with gold ions. This procedure eliminates the need to move archaeological collections, because the replicas can be made wherever the collections are held; samples can be observed in high vacuum mode, which affords a higher degree of resolution. It also makes work sessions easier, given that up to twenty-one samples can be examined in two hours. Moreover, it facilitates the analysis of pieces that because of their dimensions would not have fit into the SEM sample chamber or of modifications that as a result of their shape or position would also have presented difficulties [2].

THE STYLES OF SHELL OBJECTS IN TENOCHCA OFFERINGS

In the offerings from the Templo Mayor of Tenochtitlan more than 3000 shell elements have been found, including complete pieces as well as fragments. The materials used are species from both Atlantic and Pacific littorals. In the corpus, objects made of the pearly clam *Pinctada mazatlanica* and shell pendants made from the *Oliva* genus predominate [22]. The majority of objects were found in construction stage IVb (1469–1481), corresponding to the reign of *Axayacatl*. The low number of these pieces in earlier and later periods is noteworthy. This is perhaps due to the differential preservation of architectural periods of the sacred Tenochca precinct, caused by its destruction following the Spanish conquest.

Despite the foreign character of raw materials, what is remarkable is that many of the objects represent iconographic elements characteristic of Nahua deities from Central Mexico. This is the case of *anahuatl* pectorals, typical of Tezcatlipoca and star warrior deities; *oyohualli* pendants associated with Tlahuizcalpantecuhtli, and deities of music and dance; *yacametztli* nose ornaments often worn by moon and pulque gods, among others [23].

It is important to highlight that many of these objects were found almost exclusively in offerings at the Templo Mayor, while they were absent in nearby structures and no specimens identical in shape or raw material have been found at any other location in the Basin of Mexico. An interesting example is the *ehecacozcatl* found in Hualquila, Iztapalapa, which differs notably from Tenochca examples, because the former is made from the *Strombus gracilior* and shows perforations for suspension, while the latter are invariably made from *Turbinella angulata* and lack any holes [24-25]. Similarly, the fish made of mother-of-pearl found in Tenochtitlan and Tlatelolco display highly different shapes and raw materials, *Pinctada mazatlanica* in the case of the former and a freshwater shell (Unionidae) in the case of the latter.

Therefore, many of the shell pieces appear to be exclusive, not only to Tenochtitlan, but also to its most hermetic and elite ritual represented by the burial of offerings in the supreme temple of the Mexica empire. This strongly suggests that their manufacture must have been local and controlled by the state apparatus.

Through technological analysis it was possible to detect considerable standardization in processes and tools, which made it possible to propose the existence of a Tenochca technological style [2]. Therefore, for example, in the case of objects made from *Pinctada mazatlanica*, it was possible to determine that evidence of abrasion on surfaces and edges of objects was invariably made with basaltic rock; cuts, openwork, and incisions were made with obsidian tools; and holes were drilled with flint perforators [2].

In the case of shell pendants of the *Oliva* genus, a tendency was detected for technological standardization, in which the elimination of the shell spire was done through direct percussion, abrasion with a passive tool—which was always basalt—and a combination of both techniques. The grooved perforations in all cases were done with sharp obsidian tools. However, a few pieces were detected that were worked with materials totally different from this set, such as the cutting of the spire with obsidian powder and conical perforations made with sand. It has been proposed that these objects represent foreign production [2].

The technical analysis of manufacturing employed for the production of shell pendants not from the *Oliva* genus (*Nerita, Neritina, Cassis, Polinices, Columbella, Nitidella, Olivella, Agaronia, Marginella* and *Conus*) (Figure 1) has made it possible to identify the use of a sandstone not found in the Basin of Mexico to eliminate the spire of some shells, smooth the surface of others through abrasion, and produce irregular perforations (Figure 2). The fact that most species identified come from the Atlantic coast makes it highly likely that these objects had reached the Basin of Mexico from the Gulf Coast, perhaps from the Huasteca region, where there is an abundance of this type of sedimentary rock. This area was conquered by the Aztec Empire during the reign of Moctezuma Xocoyotzin (1440–1469), and remained subjugated until the rule of Moctezuma Ilhuicamina (1502–1520) [26-27].

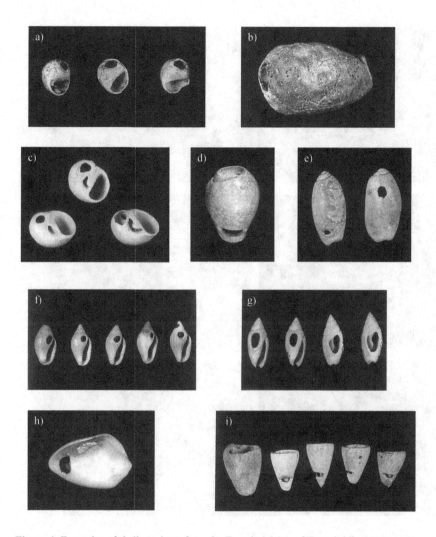

Figure 1. Examples of shell pendants from the Templo Mayor of Tenochtitlan: a) *Neritina virginea*; b) *Cassis* sp.; c) *Polinices* sp.; d) *Columbella rusticoides*; e) *Agaronia* sp.; f) *Nitidella nitida*; g) *Olivella* sp.; h) *Marginella apicina*; and i) *Conus* sp. (photographies by Germán Zúñiga Amézquita).

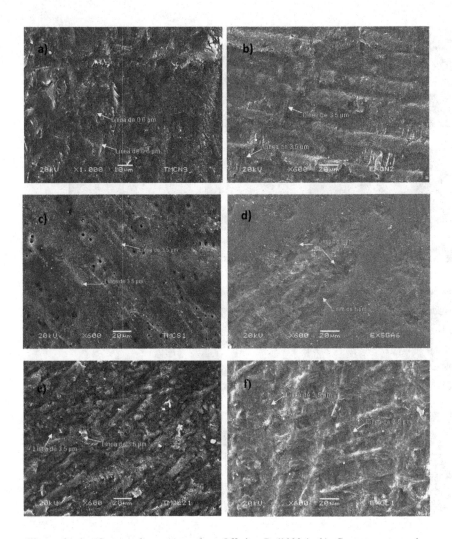

Figure 2. a) *Conus* spire cross-section, Offering P (1000x); b) *Conus spurius* spire abrasion with sandstone (600x); c) *Cassis* surface, Offering P (600x); d) *Strombus galeatus* abrasion with sandstone (600x); e) irregular perforation by abrasion on *Olivella*, Offering 60 (600x); and f) irregular perforation by abrasion produced with sandstone on *Olivella* (600x) (TMOCMP project micrographs).

Pertinent for present discussion are six *Olivella volutella* pendants, found in Burial 1 of Building 1 in Tancama, Querétaro, a Postclassic Huastec site. These objects are identical to those found in the offerings of the Templo Mayor of Tenochtitlan. The study of their manufacturing traces permitted the identification of sandstone as the type of rock employed to produce its irregular perforations through abrasion (figure 3) [26].

Figure 3. a) *Olivella volutella* pendants from Burial 1, Structure 1, Tancama (photography by Germán Zúñiga Amézquita); b) traces of the irregular perforation in one of them seen with SEM at 600x (TMOCMP project micrograph).

One of the last offerings excavated by the Templo Mayor Project, number 133, contains unique shell objects, never before found, which might also represent foreign production. This deposit was located under the south wall of the cist of Offering 126, beneath the Tlaltecuhtli goddess monolith. In a space of approximately 40 cm², immersed in a mixture of black clay, fifty-two shell elements were found. They pertained to different species, were in extraordinarily good condition, and had been covered by a layer of sea sand. This deposit corresponds to construction phase VI of the Templo Mayor, which has been attributed to *tlatoani* Axayacatl. He ruled Tenochtitlan from 1486 to 1502 and during his reign the Aztec Empire conquered parts of the Pacific coast.

The mollusk species identified correspond to the time of territorial expansion under the Triple Alliance. Among the fifty-two elements found, only one bivalve from the Caribbean has been identified (*Pinctada imbricata*) as well as one land snail.

Thirty-seven elements display evidence of human modification. What stand numerically are those that can be classified as pendants (29), which have one or more perforations for suspension that are not in the middle of the object [28]. Three correspond to gastropods: two of them are of the species *Agaronia propatula*, from which half the spire has been removed, and they have a broken outer lip; the other is of the species *Oliva julieta*, in which only the spire was removed. The remaining twenty-six elements are bivalves of the species *Cardita crassicostata* (1), *C. megasthropa* (1), *Chione subrugosa* (1), *C. undatella* (10), *Glycymeris gigantea* (8), and *G. multicostata* (5) (figure 4). It is worth noting that for the production of pieces, young specimens of the smallest sizes were chosen, ranging from 1.4 to 2.3 cm in length, 1.2 to 2.7 cm in width and 0.4 to 0.8 cm in thickness. In all cases, the outer surfaces of the shells have a remarkable sheen, the result of polishing and burnishing them.

Twenty-five pendants are complete; eighteen of them display abrasion on the umbone and six on the hinge only one case has both of these modifications. Nine have irregular perforations by abrasion on the umbo, while thirteen have conical drilling, seven on the umbo and six on the hinge. Only two pieces have tubular perforations on the umbo. Only one of the objects has two perforations, an unsuccessful conical one on the hinge, and the other an irregular hole produced by abrasion on the umbo. The remaining pendant is a half valve, which was cut in half lengthwise and it is possible that it might have been a reused piece, given that the umbo is eroded.

Eight *Polinices uber* shells can only be classified as worked gastropods, given that their outer surface was polished and burnished and their outer lip was abraded (figure 4). They do not have perforations of any type. In this case as well, only young specimens were chosen that do not exceed 3.5 cm in length, 2.9 cm in width, and 1.9 cm in height.

Analysis of the manufacturing traces made it possible to identify flint as the material for the tool to abrade the umbos and the outer surfaces of shells, and also for polishing; it was also used to cut the spires of the shells and to produce conical perforations. Flint is a material that produces traces of workmanship in the form of fine lines, between 0.6 and 2 µm in width, which can appear as parallel lines, forming bands of larger dimensions in which micro lines can be seen; or else they can appear as bands between 4 to 5 µm in width, which crisscross to produce rough surfaces. Polishing with flint produces straight lines of the dimensions mentioned earlier, which go in different directions and crisscross (figure 5).

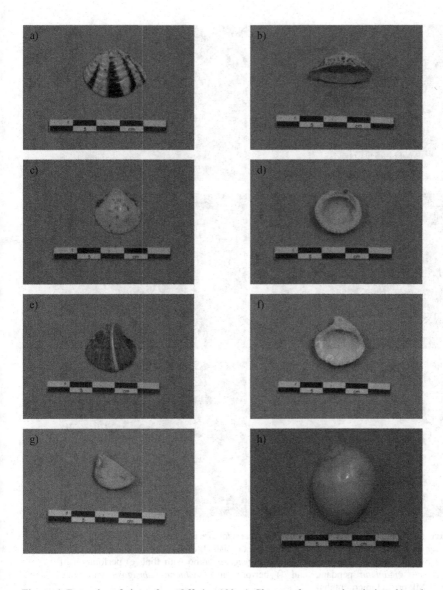

Figure 4. Examples of pieces from Offering 133: a) *Chione subrugosa*, dorsal view; b) and of the umbo; c) *Glycymeris multicostata*, dorsal view and (d) ventral view; e) *Cardita megastropha*, dorsal view and f) ventral view; g) *Chione undatella* cut in half; and h) *Polinices uber* (photos by Germán Zúñiga Amézquita).

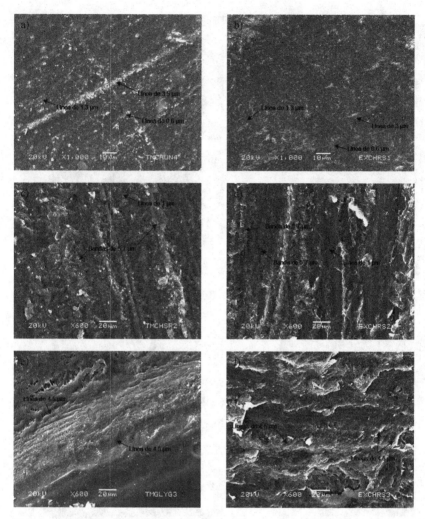

Figure 5. SEM images: a) surface of *Chione undatella* pendant; b) surface of *Chione subrugosa* abraded and polished with flint, and burnished; c) abrasion of umbo of *Chione subrugosa* pendant; d) abrasion of *Chione subrugosa* umbo with flint; e) perforation of *Glycymeris gigantean* pendant; and f) perforation of *Chione subrugosa* with flint (TMOCMP project micrographs).

The use of burnishing could be corroborated experimentally by polishing a *Glycymeris* sp. valve with a flint nodule for 2 hours and 56 minutes, which did not produce a sheen similar to the archaeological specimens. Later the piece was burnished with leather for 23 minutes, which produced a lustrous finish similar to the archaeological pieces.

DISCUSSION

The analysis of the manufacturing traces with SEM made it possible to identify with a fairly high degree of certainty the materials and processes employed to manufacture shell objects from the pre-Hispanic period. As a result, the decisions made by ancient artisans in the operational chains have been inferred, which has served as the basis to propose technological styles. Obviously this type of analysis must be complemented by the study of shapes and decoration of pieces, which as a whole permit more comprehensive discussion of the origin of manufactures.

In the case of the pieces found in Tenochca offerings, earlier the existence of a group of objects of local production had been proposed, which took place in the context of the Mexica palace. Part of this production was intended for the most exclusive and elite religious practices in Mexica society. Similarly, some pieces have been detected, mainly shell pendants, whose production tools indicate foreign production, possibly on the Gulf Coast of Mexico, specifically the Huasteca region.

The discovery of Offering 133 has made it possible to become familiar with another group of objects, different from earlier specimens, both in form and in technique of manufacture. Remarkably, all of these species are from the Pacific coast and the burial of the offering dates to the reign of Ahuizotl, when the Mexicas conquered parts of the Pacific coast. It can be posited with some certainty that these pieces are of foreign production, because they do not correspond to techniques known for Tenochca objects; however, unfortunately it is not possible to suggest a more specific zone where these objects might have been manufactured, because no similar specimens from other sites have been found to date. As the interest in studying collections of shell objects grows, it is likely that these sorts of questions will be resolved.

REFERENCES

1. L. López Luján, *Las ofrendas del Templo Mayor de Tenochtitlan*, Mexico City, Instituto Nacional de Antropología e Historia, INAH, Mexico, 1993.
2. A. Velázquez Castro. *La producción especializada de los objetos de concha del Templo Mayor de Tenochtitlan* Instituto Nacional de Antropología e Historia INAH, Colección Científica 519, Mexico, 2007.
3. E. Melgar Tísoc, Tradiciones tecnológicas en la lapidaria de obsidiana en el Templo Mayor de Tenochtitlan, in *Producción artesanal y especializada en Mesoamérica*, Linda R. Manzanilla and Kenneth G. Hirth (eds.) Instituto Nacional de Antropología e Historia INAH - Universidad Nacional Autónoma de México, UNAM, Mexico, 2010, 205–226.
4. N. Schulze, *El proceso de producción metalurgia en su contexto cultural: los cascabeles de cobre del Templo Mayor de Tenochtitlan,*" Ph.D. dissertation in Anthropology, Facultad de Filosofía y Letras, , Instituto de Investigaciones Antropológicas, Universidad Nacional Autónoma de México, Mexico, 2008.
5. A. Velázquez Castro, *Ancient Mesoamerica*, 22 (2011) 437-448.

6. P.G. Roe, 1995 Style, society, myth and structure, in: *Style, Society and Person*, Christopher Carr and Jill E. Neitzel (coords.), Plenum Press, New York, 1995, 27–76.

7. Ch. Carr, Building a unified middle range theory on artifact design: historical perspective and tactics in: *Style, Society and Person*, Ch. Carr and J.E. Neitzel (coords.), Plenum Press, New York, 1995, 151–169.

8. M. Shanks, Ch. Tilley, Material culture, in *Social Theory and Archaeology*, M. Shanks and Ch. Tilley (coords.), University of New Mexico Press, Albuquerque, 1987, 88–94.

9. M. Shanks, Ch. Tilley, *Re-Constructing Archaeology, Theory and Practice*, Routledge, London, 1994.

10. J.E. Clark, W.J. Parry, Craft, *Research in Economic Anthropology* 12, 1990, 289–346.

11. Ch. Carr, J.E. Neitzel, High level theory on the causes of style, in: *Style, Society and Person*, Ch. Carr and J.E. Neitzel (coords.), New York, Plenum Press, 1995, 21–25.

12. J.A. Román Berrelleza, L. López Luján, *Arqueología Mexicana* vol. 8, no. 40 (1999).

13. X. Chávez Balderas, Rituales funerarios en el Templo Mayor de Tenochtitlan, Dissertation thesis in Archaeology, Mexico City, Escuela Nacional de Antropología e Historia, Instituto Nacional de Antropologia e Historia, INAH, Mexico 2002.

14. P. Lemonnier, *Journal of Anthropological Archaeology* 5 (1986) 147–186.

15. B. Pfaffenberger, *Man* 23, no. 2 (1988) 236–252.

16. M.B. Schiffer, *Technological Perspectives on Behavioral Change*, University of Arizona Press, Tucson, 1992.

17. A. Leroi-Gourhan, *L'homme et la matière*, Éditions Albin Michel, Paris, 1943.

18. A. Leroi-Gourhan, *Milieu et techniques*, Éditions Albin Michel, Paris, 1945.

19. O.P. Gosselain, *Man* vol. 27, no. 3 (1992) 559–583.

20. J.R. Sackett, Style and ethnicity in archaeology: the cause of isochrestism, in: *The Uses of Style in Archaeology*, M. Conkey and Ch. Hastorf (coords.), Cambridge University Press, Cambridge, Massachusetts, 1990, 32–43.

21, M.T. Stark, Social dimensions of technical choice in Kalinga ceramic tradition, in: *Material Meanings*, E.S. Chilton (coord.), The University of Utah Press, Salt Lake City, 1999, 24–44.

22. A. Velázquez Castro, *Tipologia de los objetos de concha del Templo Mayor de Tenochtitlan*, Instituto Nacional de Antropología e Historia (INAH), Colección Científica no. 392, 1999, Mexico.

23. A. Velázquez Castro, *El simbolismo de los objetos de concha encontrados en las ofrendas del Templo Mayor de Tenochtitlan*, Instituto Nacional de Antropología e Historia (INAH), Colección Científica no. 403, 2000, Mexico.

24. E. Mancha González, Objetos de concha en contextos arqueológicos de la Cuenca de México, en la época prehispánica Dissertation thesis in Archaeology, Mexico City, Escuela Nacional de Antropología e Historia, Instituto Nacional de Antropologia e Historia, INAH, Mexico 2002.

25. A.Velázquez Castro, E. Melgar, La elaboración de los ehecacozcatl de concha del Templo Mayor de Tenochtitlan, in: *Arqueología e historia del Centro de México, Homenaje a Eduardo Matos Moctezuma*, L. López Luján, D. Carrasco and L. Cué, (eds.), Mexico City, Instituto Nacional de Antropología e Historia, 2006, 525–537.

26. A.Velázquez Castro, B. Zúñiga Arellano, Á. González López, *Nerita* Shell Objects in the Offerings of the Great Temple of Tenochtitlan, in *2nd Latin American Symposium on Physical and Chemical Methods in Archaeology, Art and Cultural Heritage Conservation & Archaeological and Arts Issues in Material Science – IMRC 2009*, J. L.

Ruvalcaba Sil, J. Reyes Trujeque, J. A. Arenas Alatorre and A. Velázquez Castro (eds.), Universidad Nacional Autónoma de México, Universidad Autónoma de Campeche and Instituto Nacional de Antropología e Historia,Mexico, 2010, 107–111.

27. A. Velázquez Castro, Adrián, B. Zúñiga Arellano, *Pendientes de caracol de las ofrendas del Templo Mayor*, paper presented at the Congreso Internacional "Culturas Americanas y su Ambiente: Perspectivas desde la Zooarqueología, Paleobotánica y Etnobiología," Mérida, Yucatán, 2010.

28. Ma. de L. Suárez Diez, *Tipología de los objetos prehispánicos de concha,* Instituto Nacional de Antropología e Historia (INAH), Colección Científica no. 403, 1977, Mexico.

Mater. Res. Soc. Symp. Proc. Vol. 1374 © 2012 Materials Research Society
DOI: 10.1557/opl.2012.1381

Technological and Material Characterization of Lapidary Artifacts from Tamtoc Archaeological Site, Mexico

Emiliano Melgar[1], Reyna Solís[1], José Luis Ruvalcaba[2]
[1]Museo del Templo Mayor, Instituto Nacional de Antropologia e Historia, INAH.
Seminario No. 8, Centro Histórico, Mexico DF 06060, Mexico.
e-mail: anubismarino@gmail.com
[2]Instituto de Física, Universidad Nacional Autónoma de México, UNAM.
Apdo. Postal 20-364, Mexico DF 01000, Mexico. e-mail: sil@fisica.unam.mx

ABSTRACT

Tamtoc is a very important archaeological site in San Luis Potosi, in the Central region of Mexico. The pre-Hispanic Huastec culture developed in this site (900-1100 A.D.). During the archaeological excavations, a large amount of lithic artifacts were recovered from burials and offerings. Among them, pieces of semitransparent crystalline objects of color blue, green, yellow and white and green stone pieces were discovered in one of the most important ceremonial precinct, inside a water reservoir of the monument 32 "The Priestess". The aim of this work is to measure the composition of the artifacts for provenance study and to establish the manufacturing technique and tools used to produce them. For material analysis, a combined analysis involving X-ray Fluorescence (XRF) and Raman spectroscopy was applied. The main elements as well as some traces can be measured by XRF while the mineral identification can be established by Raman. The results indicate that most of the pieces are calcite with traces of rare elements.

On the other hand, experimental archaeology using a well established methodology of optical and electron microscopy examination of the manufacturing traces was applied. From this technological study the specific use of tools and materials were established for this site.

INTRODUCTION

During the archaeological excavations at Tamtoc, an archaeological site of the Huastec people (900-1100 A.D.) in San Luis Potosi, northern México, hundreds of polished stones were recovered in offerings and burials. Among them, the context with the highest number of pieces is the spring water associated with the famous Monument 32 known as "The Priestess" (Figure 1). There, a huge amount of translucent beads and pendants of four colors (blue, green, yellow and white) were deposited inside the water as an offering representing the "petrified waters" for rituals of fertility (Figure 2). Also, in this context there are evidences of their production, like unfinished pieces and fragments without any modification of these raw materials. In contrast, all the burials have dark green objects employed as prestige goods to emphasize their power and status (Figure 3). It is interesting that we do not have any evidences of production of these pieces.

Because of that, the aim of this work is to measure the composition of these artifacts to determine their provenance and to establish the manufacturing technique and tools used to produce them. For material analysis, a combined analysis involving X-ray Fluorescence (XRF) and Raman spectroscopy was applied. For the technological analyses we employed experimental archaeology using a well established methodology of optical and electron microscopy examination of the manufacturing traces.

Figure 1. The Monument 32 or "The Priestess" and the associated spring water at Tamtoc.

Figure 2. The four colors (blue, yellow, white and green) of the translucent stones at Tamtoc.

Figure 3. Some dark green pendants associated with a burial at Tamtoc.

METHODOLOGY

In this study we analyzed 296 pieces (Figures 4 and 5): 47 beads, 68 pendants, 1 earplug, 126 unworked pieces, 14 unfinished objects, 1 debitage, 14 failed pieces, and 2 recycled objects.

In order to know what raw materials were used for these pieces, where are located their sourcing areas, and which tools were employed for manufacture the beads and pendants, we applied a set of non-destructive analyses:

The chemical characterization of the raw materials was done at the Physics Institute of UNAM, using X-Ray Fluorescence (XRF) and Raman Spectroscopy (Raman). The first one, Sandra equipment (Figure 6a), employed an X-ray beam of Molybdenum (Mo), obtaining a detailed analysis of the elements of the materials. The second one, a Raman Inspector Delta Nu equipment (Figure 6b), employed a Red Laser Beam (785nm) that illuminated the sample, the light emitted is characteristic of the minerals in the analyzed sample. These analyses do not require sampling the pieces, because they do not touch them. Also, they are fast because we require some minutes to obtain results. Finally, the combination of both techniques improved the results obtained [1,2].

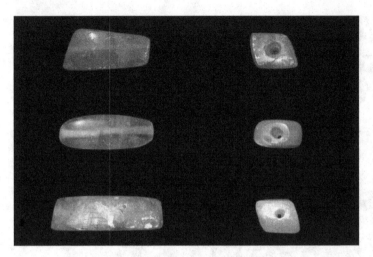

Figure 4. Examples of objects made with the translucent stones.

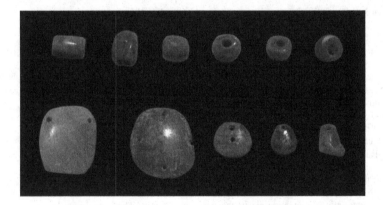

Figure 5. Examples of objects made with the dark green stones.

Figure 6. The equipments employed in the analysis: XRF (a) and Raman (b).

The technological analysis was done at the Experimental Workshop on Lapidary Objects in the Great Temple Museum and the SEM Lab of INAH. The manufacturing traces obtained in the experiments (Figure 7), using tools and techniques reported in the historical sources and the archaeological contexts [3,4,5,6], were characterized and compared with the archaeological ones, using an Optic Microscopy (OM) at 10x and 30x, and Scanning Electron Microscopy (SEM), at 100x, 300x, 600x, and 1000x. The last one was used with the same parameters proposed by Velázquez Castro [7]: High Vacuum Mode (HV), 20 kV of energy, SEI signal, spotsize of 42 and 10mm of work distance.

Figure 7. Experimental archaeology on lapidary objects: abrading of calcite with sandstone.

RESULTS

Chemical analyses

Elemental analyses of the translucent objects with XRF showed that their principal element is Calcium (Ca) with minor concentrations (<1%) of iron (Fe) and Strontium (Sr), and the phosphorus (P) is absent (Figure 8a). Only in the green pieces the iron is higher. And with this technique it is not possible to detect Fluorine (F) or Magnesium (Mg). With Raman, these pieces showed spectra of Calcite and Calcium Carbonate with signals of 280 and 1088 cm^{-1} (Figure 8b). They are quite similar in the case of white and yellow pieces, but in the case of the green ones, they have other minerals, like Cerium (Ce) and Yttrium (Y). So, these pieces are calcites, probably from the Tanchipa Mountains near Tamtoc, the closest mines of calcite in the region.

Elemental analyses of the dark green objects with XRF showed that they have Silicon (Si), Potassium (K), Iron (Fe), Nickel (Ni), Chromium (Cr), and Manganese (Mn), among others (Figure 8a). With Raman, these pieces showed spectra of Quartz (463 cm^{-1}), Magnesite (1090 y 328 cm^{-1}), and Aluminum silicates of the Serpentine Family (Figure 9). So, these pieces are green quartzites, probably from the Southern Mountain Range of Mexico and Guatemala.

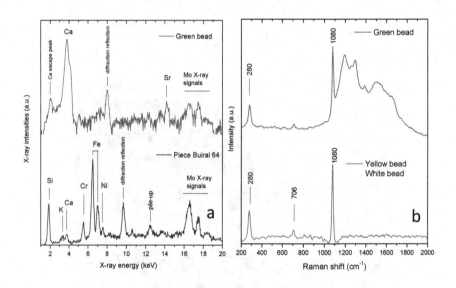

Figure 8. Analysis of the translucent and dark green objects with XRF (a) and Raman (b).

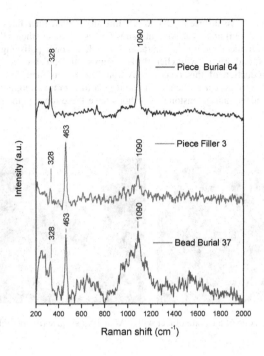

Figure 9. Raman analysis of the dark green objects.

Technological analyses

With the Optic Microscopy, the pieces showed two patterns: The calcite objects have straight lines mixed with luster areas (Figure 10a) and the drill holes present soft circular striations (Figure 10b).

In contrast, the quartzite pieces have high polished surfaces (Figure 11a) with very fine lines and their holes present clear circular striations (Figure 11b).

With SEM, the pieces confirm these two patterns: The calcite pieces have flat bands of 100 μm on surfaces (Figure 12a) and diffused lines of 1 μm inside the drill holes (Figure 13a). This matches with the experimental abrading with basalt, brightening with leather, and drilling with chert powder and reed (Figures 12b and 13b). Because the evidences of production of the calcite pieces share this technology with the finished objects, and the tools identified (basalt and chert) were found in this site, it could be a local manufacturing tradition.

The quartzite objects have diffused bands of 20 μm mixed with finer lines of 0.6 μm on surfaces (Figures 14a and 15a) and clear straight bands of 2-4 μm inside the drill holes (Figure 16a). This matches with the experimental abrading with limestone, polishing with jadeite, brightening with leather, and drilling with chert drills (Figures 14b, 15b and 16b). Because there is no evidences of production of this pieces, only finished objects, and some of the tools (limestone and jadeite) were absent in the site, it could be a foreign tradition, perhaps from the Maya area where the jadeite and greenstone workshops have these tools as their technological characteristic.

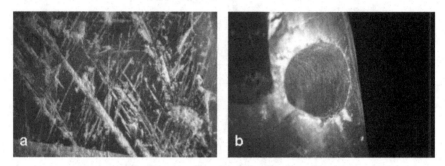

Figure 10. Analysis of a calcite object with OM at 10x: surface (a) and hole (b).

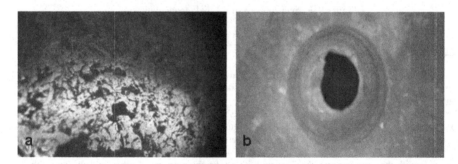

Figure 11. Analysis of a quartzite object with OM at 10x: surface (a) and hole (b).

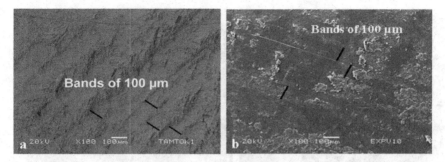

Figure 12. Analysis of surfaces with SEM at 100x: calcite object (a) and experimental abrading with basalt (b).

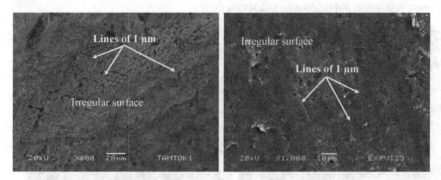

Figure 13. Analysis of holes with SEM at 1000x: calcite object (a) and experimental drilling with chert powder and reed (b).

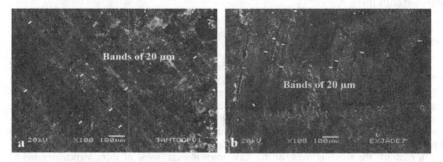

Figure 14. Analysis of surfaces with SEM at 100x: quartzite object (a) and experimental abrading with limestone (b).

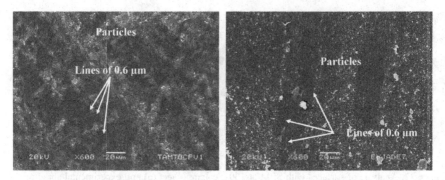

Figure 15. Analysis of surfaces with SEM at 1000x: quartzite object (a) and experimental polishing with jadeite and brightening with leather (b).

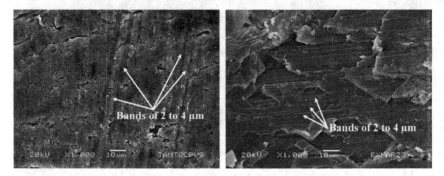

Figure 16. Analysis of holes with SEM at 1000x: quartzite object (a) and experimental drilling with chert drills (b).

DISCUSSION

The results obtained allowed us to appreciate two patterns of raw materials and technological traditions. But, are they local or foreign?

In the case of the calcites, despite their color (green, white yellow and blue), there are some mines and sourcing areas at Tanchipa Mountains near Tamtoc. Also, there are a lot of evidences of production of the beads and pendants made with these raw materials, and the tools identified were found in the site. Because of that, we infer that these objects were produced locally and almost all of the manufacturing process took place in the lapidary workshops at Tamtoc.

In contrast, the sourcing areas of the dark green objects, some of them serpentines and quartzites, are at the Southern Mountain Range of Mexico and Guatemala [8], hundreds of kilometers away from Tamtoc. Perhaps they come from Chiapas and Guatemala because the present the same tools and techniques of the jadeite and greenstone workshops of the Maya area

[4, 5], and these tools are absent at Tamtoc. Also, there are no evidences of production of these pieces in the site, and all of them are finished objects deposited inside the burials as offerings and prestige goods. Based on this info, we infer that these objects are foreign manufactures, perhaps Mayan pieces.

CONCLUSIONS

The combination of the chemical and technological analyses allowed us to identify the raw materials and the tools employed by the ancient people of Tamtoc. Also, we could distinguish two technological traditions related with the provenance of the raw materials and the tools employed.

One tradition is characterized by the calcite objects and their evidences of production as an offering in the spring water associated with the Monument 32. These pieces could come from the Tanchipa mountains near Tamtoc and the tools identified in the technological analysis were found in the site (basalt metates and chert flakes). Because of that, we infer that these objects were produced locally and almost all of the manufacturing process took place in the lapidary workshops at Tamtoc.

The other tradition is characterized by the quartzite and serpentine pieces found in the burials. They could come from the Southern Mountain Range of Mexico and Guatemala. Also, they could arrive at Tamtoc already manufacture, perhaps from the Maya Area, because the tools identified (limestone metates and jadeite polishers) are absent and they match with the technological tradition of the Mayan lapidary workshops.

ACKNOWLEDGMENTS

The authors would like to thank Mtra. Estela Martínez and Mtro. Guillermo Córdova, directors of the Tamtoc Archaeological Project: Origin and Development of the Urban Landscape of Tamtoc, for the permissions to analyze the lapidary objects and for their assistance of the contextual information of the site. Also, we thank the members of the Experimental Workshop on Lapidary Objects (Mauricio Valencia, Isaac Ramírez, Mijaely Castañón, and Hervé Monterrosa) for the experiments, and Mtro. Gerardo Villa for obtaining the images with the SEM at the access to the SEM Lab of INAH.

Authors acknowledge the supports of UNAM PAPIIT IN403210 project as well as the CONACYT Mexico grant 131944 to carry out this study. This research has been performed in the frame of the Non Destructive Study of the Mexican Cultural Heritage ANDREAH network (www.fisica.unam.mx/andreah).

REFERENCES

1. J. L. Ruvalcaba, L. Filloy, M. Vaggli, L. Tapia, R. Sánchez, Estudio no destructivo *in situ* de la Máscara de Malinaltepec, in: *La Máscara de Malinaltepec*, ed. S. Martínez, Instituto Nacional de Antropología e Historia, INAH, Mexico, 2010, 153-168.
2. J.L. Ruvalcaba, D. Ramírez, V. Aguilar and F. Picazo. *X-Ray Spectrometry*. 39 (2010) 338-345.
3. K. Aoyama. *Latin American Antiquity* 18 (2007) 3-26.

4. B. Kovacevich. Ritual, Crafting, and Agency at the Classic Maya Kingdom of Cancuen, in: *Mesoamerican Ritual Economy. Archaeological and Ethnological Perspectives*, E.C. Wells and K. L. Davis, Boulder eds, 2007, 67-114.
5. E. Rochette. *Archaeological Papers of the AAA* 19 (2009) 205-224
6. F. B. Sahagún. *Historia General de las Cosas de Nueva España*, Ed. Porrúa, Mexico, 1956.
7. A. Velázquez Castro, *La producción especializada de los objetos de concha del Templo Mayor de Tenochtitlan*. Instituto Nacional de Antroplogía e Historia INAH, Colección Científica 519, Mexico, 2007.
8. J. Robles and A. Oliveros. *Arqueología* 35 (2005) 5-22.

Mater. Res. Soc. Symp. Proc. Vol. 1374 © 2012 Materials Research Society
DOI: 10.1557/opl.2012.1382

Comparative Study of Two Blue Pigments from the Maya Region of Yucatan

Silvia Fernández-Sabido[1], Yoly Palomo-Carrillo[2], Rafael Burgos-Villanueva[2] and Romeo de Coss[1]

[1] Department of Applied Physics, Cinvestav-Mérida, A.P. 73 Cordemex, C.P. 97310, Mérida, Yucatán, Mexico. e-mail: silviafidelina@hotmail.com
[2] Department of Archaeology, Centro INAH Yucatán, Antigua carretera a Progreso km. 6.5, S/N, C.P. 97310, Mérida, Yucatán, Mexico.

ABSTRACT

A comparative study of two blue pigment found in separate megalithic structures in *Yucatán México* is presented. The first sample (M1) is a piece of turquoise stucco discovered at the top of a building known as Structure-2 in the town of *Dzilam González*. The second sample (M2) is a residual blue powder that was contained in a *Oxcum Café* type ceramic vessel recovered in the rubble of the *Kabul* building in *Izamal* city. The interest in characterizing these samples increases with the possibility of finding in them evidence of Maya Blue, a dye created in the eighth century by the Maya people, whose extraordinary physical and chemical properties have been studied in laboratories around the world. Maya Blue was a tailored technology used for several centuries, even during the Spanish occupation, throughout Mesoamerica. Despite 80 years of study, the mysteries of its composition, traditional preparation and obsolescence have not yet been fully resolved. Using different spectroscopic techniques (SEM, EDX, XRD, FTIR, UV-Vis DR) we have studied and compared the blue colorants in M1 and M2. Results indicate that M1 is Maya Blue. Despite some similarities in the infrared vibrational spectra of the two samples, we have determined that M2 is not Maya Blue but a non-Mesoamerican mineral pigment known as Ultramarine which was probably introduced to America by Europeans.

INTRODUCTION

Maya Blue is a colorant of extraordinary physical and chemical properties that has been found in different types of objects and structures of ancient Mesoamerica [1-5]. Archaeological evidence points the Mayans of eighth century as the inventors of this particular technology [3,6,7] and since its re-discovery in *Chichén Itzá* by Merwin [8] almost 80 years ago, the Maya Blue has been studied at laboratories around the world. At present, it is generally accepted that this pigment is a hybrid material composed of a mineral part formed by the palygorskite clay [1, 2, 4, 9, 10] in whose nanometric channels is contained the organic component believed responsible for the color: the indigo [2-4, 11]. However, the lack of historical documents that describe the method by which the ancient Maya preparing this dye [12], in addition to the fact that is not yet been possible to separate the organic component from the inorganic matrix in studied samples, it still can hypothesize about the presence of one or more organic components than the indigo. For all these reasons, the study of blue dyes collected from archaeological contexts in the Maya area is relevant to identifying potential prehispanic-precursors or colonial-replacements of Maya Blue.

EXPERIMENT

The M1 sample is a piece of turquoise stucco (figure 1.a) with an approximate size of 4 x 3 cm found in July 2010 at the top of a building known as Structure-2 (figure 1.b) in the town of *Dzilam González*. This building has a timeline that runs from the Upper Preclassic (350 - 150 BC) to the Early Classic period (250 - 600 AD). On the other hand, M2 is a 39 mg of residual deep-blue powder contained in a *Oxcum Café* type ceramic vessel (figure 1.c) recovered in April 2010 in the rubble of the *Kabul* building (figure 1.d) in *Izamal* city. Although this architecture is dated to the Early Classic, the vessel is dated to the Colonial period (1550 - 1800 AD). Both localities are in the central region of the *Yucatán* Peninsula in *México*.

Figure 1. (a) M1 stucco sample; (b) Structure-2 in *Dzilam González* (c) M2 residual powder in a *Oxcum Café* type ceramic vessel; (d) *Kabul* building in *Izamal*.

The external morphology and elemental composition for these samples, were analyzed by Scanning Electron Microscopy (SEM) and Energy Dispersive X-Ray Spectroscopy (EDX) in a XL30 ESEM Philips equipment. For avoid scanning faults due to charge, a ultrathin gold/palladium coating were sputter on the samples. The phase composition was determined by X-ray Diffraction (XRD) on a Siemens D5000 diffractometer and on an ItalStructures High Resolution X-ray Diffractometer (HRD) 3000 with CuKa1 monochromatic radiation. The pre-treatment for carbonate elimination (HCl 5%) was applied. The molecular chemical bonds were studied by Fourier Transform Infrared Spectroscopy (FTIR) in a Thermo Nicolet Nexus 870 with 32 scans and 4 cm^{-1} of resolution; 5 mg of the samples have been milled with 195 mg of potassium bromide (KBr). The Diffuse Reflectance (DR) spectra were recorded using an ultra violet–visible (UV–VIS) AvaSpec-2048 optical fiber spectrometer with integrating-sphere setup and an AvaLight-DH-S-BAL Deuterium/Halogen ligth source.

RESULTS

External morphology

Figure 2 shows SEM images of a pigment layer of M1 stucco and the M2 powder pigment. We clearly note in M1 (figure 2.a-b) the presence of a fibrous material similar to clay palygorskite found in Maya Blue samples. This fibrous structure is not present in M2 pigment (figure 2.c-d).

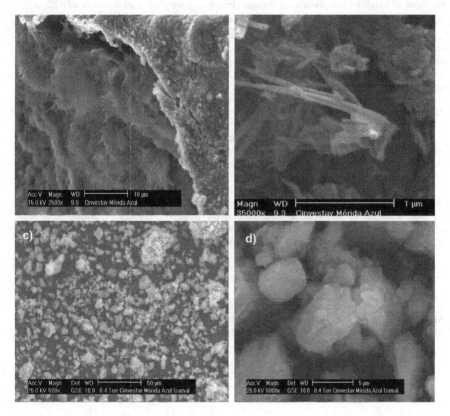

Figure 2. SEM images for: (a-b) M1 stucco pigment; (c-d) M2 residual blue powder.

Composition

The elemental composition obtained by EDX and the XRD spectra are presented in Table I and Figure 3, respectively. For XRD analysis, natural palygorskite collected in *Yucatán* Peninsula was used as a reference and a pre-treatment with HCl 5% was applied to samples for carbonate elimination. The results indicate that M1 contains palygorskite, the major component of Maya Blue. The theoretical formule of this clay is $Si_8O_{20}Al_2Mg_2(OH)_2(H_2O)_44(H_2O)$ [5]. They also seem to be present other minerals as clinoptilolite or heulandite.

In the other hand, the M2 sample, analyzed before HCl treatment, contains a mineral dye known as Ultramarine Blue, and calcite comming from the vessel or the ground. Ultramarine was an expensive pigment extracted from the semi-precious stone Lapis Lazuli and it was first prepared synthetically in 1828. Its theoretical formule is $Na_{8-x}[SiAlO_4]_6.[S_2,S_3,SO_4,Cl]_{2-x}$ consisting of a sodium aluminosilicate structure based on a tetrahedral arrangement of AlO_4 and SiO_4 units in a sodalite cage containing the chromophores S_3^- and S_2^- [13]. Upon contact with acid, the blue color of M2 disappeared and the XRD spectrum of brown residue shows the presence of clays or zeolites among which may include palygorskite, illite, kaolinite, montmorillonite, heulandite. This is not surprising since Ultramarine Blue may be prepared with a mixture of clays mainly kaolinite [14, 15]. An Ultramarine Blue bleaching has been reported in the literature but with stronger acids [16].

Molecular bonds

Figure 4 shows the FTIR spectra of the archaeological samples and natural palygorskite. One can observe a surprising similarity between M1 and M2 which can be explained if we take into account that in both, Maya Blue and Ultramarine, there is an aluminosilicate matrix which protects the element that gives color.

Calcite is detected by the band at 1450 cm^{-1} and the pics at 876 cm^{-1} and 712 cm^{-1} [17, 18], which are absentes in pure palygorskite. The width band from 3000 to 3500 cm^{-1} corresponds to OH from interlaminar water in aluminosilicates; the peak in 3624 cm^{-1} belonging to the hydroxyl groups of the silicates [19]. The partially resolved peaks at 1655 and 1630 cm^{-1} correspond to bending modes of absorbed and zeolitic water [17].

Table I. EDX analysis (weight %) for M1 stucco pigment and M2 blue powder.

	C	O	Na	Mg	Al	Si	S	K	Ca	Ti	Fe
M1	65.65	21.44	0.40	1.60	2.59	6.68	0.14	0.15	0.26	0.08	1.00
M2	42.25	31.35	3.08	0.00	4.89	6.25	2.96	2.56	5.70	0.00	0.98

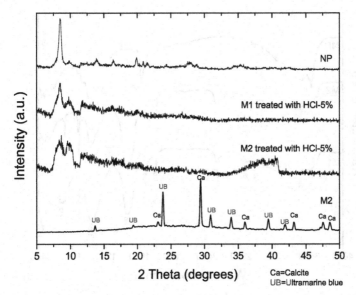

Figure 3. XRD spectra of M1 stucco pigment, M2 blue powder, and natural palygorskite (NP).

Between 1200 and 400 cm^{-1}, characteristic bands of silicate can be observed, mainly those corresponding to Si–O bonds in the tetrahedral sheet. This interval of wavenumber is very complex because the lattice modes also have an influence in this region of the spectra [17]. The intense peak near to 1030 cm^{-1} corresponding to stretching of the Si–O bond [20].

The bands between 450-550 cm^{-1} are combinations of Si-O-Al in which the vibrations are bending [21] and those between 986 and 655 cm^{-1} to M–OH deformation [17]. The peak at 912 cm^{-1} in M1 corresponds to Al–Al–OH deformation and it is a consequence of the dominantly dioctahedral character of palygorskite [22].

The absorption bands in the range of 1500-3000 cm^{-1}, where there are no characteristic bands of minerals but of organic compounds, suggesting the use of some organic medium in the preparation or the application of pigments. The band of 1800 cm^{-1} belongs to aromatic compounds and those of 2870 and 2960 cm^{-1} to CH and CH$_2$ groups, respectively.

We must take into account that the use of FTIR spectra is difficult when dealing with mixtures of materials; however, it is helpful to look at different water binding states, hydrogen bonds and OH-units.

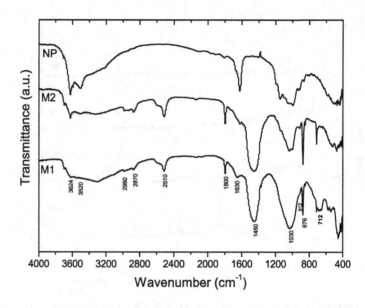

Figure 4. FTIR spectra for the M1 and M2 pigment samples compared with natural palygorskita (NP).

Difusse reflectance

The UV–Vis reflectance spectra of M1 and M2 samples are shown in figure 5. The intense fundamental vibration for M1 occurs in the visible around 660 nm which agrees with the reported results for archaeological Maya Blue, which presumely involves one molecular indigoid species, and for artificial Maya Blue reconstituted from palygorskite and indigo [23]. A strong absorption band centered at 600 nm is observed for M2, as is reported for Ultramarine Blue [24]. This is due to the charge transfer transition inside the S_3^- group that is present in the lattice of the complex aluminosilicate.

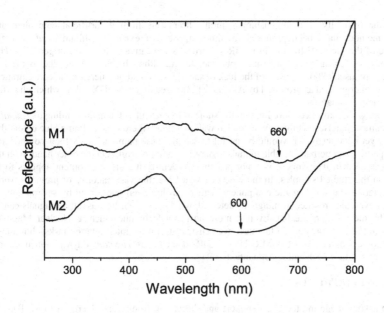

Figure 5. Reflectance spectra for M1 and M2 pigment samples.

DISCUSSION AND CONCLUSIONS

Archaeological samples of two blue pigments from *Yucatán* Peninsula in *México* were studied. We have found that the M1 stucco sample of *Dzilam González* is Maya Blue. In contrast, we have determined that the M2 blue powder of *Izamal* is not Maya Blue but a non-Mesoamerican mineral pigment known as Ultramarine. This last result opens a question: why a foreign pigment was in the rubble of a maya building? To resolve this issue would be useful to know whether the Ultramarine pigment in M2 is natural or synthetic.

The natural Ultramarine is mainly extracted from the Lapis Lazuli mines in Afghanistan since ancient times until today. This pigment was extensively used in Europe in manuscripts and paintings during the 14th and 15th centuries [25]. In addition to asian deposits, there are Lapis Lazuli mines in Chile (very far from the Mesoamerican region) which produce an Ultramarine with a paler hue [26]. Less important sources are in Russia, the United States and Italy [27]. On the other hand, the artificial Ultramarine was firstly obtained in Europe in 19th century.

For the historical contact between America and Europe, one would think that the Ultramarine was introduced to Mesoamerica by Europeans. Being that it was very expensive, the use of Ultramarine was extended when it was possible to achieve the synthetic version at low cost. For this reason it is also logical to conclude that M2 is synthetic. Unfortunately, it is already known the difficulty of distinguishing natural Ultramarine from its artificial equivalent and several methods have been proposed for this purpose. For exemple, Miliani *et al.* note the presence of the pic at 2340 cm^{-1} in the IR spectrum as indicative of a natural origen [27]. This peak does not appear to be present in our sample M2 although this may be due to the low concentration used (5%). Because of the lack of sample, we could not increase the concentration. There are other criteria as proposed by Ajò *et al.* [28] based on Na/S EDX value, which classifies the M2 sample as natural.

Future work may shed more light on the subject. In particular, the main building of *Izamal* is a 16th century franciscan convent with ancien wall paintings where we can find a blue colors that have not yet been analyzed. Would be useful to sample these paints to study their composition and properties, and perhaps be able to make a parallel with this work. On the other hand, with the techniques used in this work it was not possible to identify the organic content appears to be present in the studied samples. In the case of Maya Blue, the vast majority of previous works point indigo as the sole organic component. This result has been questioned by several authors [12]. Not even the presence of indigo, as detected by FTIR, is clear because its IR signals can be hidden by the palygorskite [4]. Even if more powerful techniques such as Nuclear Magnetic Resonance (NMR) may help in this task, we consider useful for interpretation tasks, building a spectroscopic (Raman, FTIR, NMR, UV-vis DR) database of *Yucatán* Maya region materials likely to be found in archaeological contexts.

ACKNOWLEDGMENTS

We acknowledge the technical support and discussion from Dora Huerta in SEM; Patricia Quintana, Santiago González and Daniel Aguilar in XRD; Antonio Azamar and Ricardo Mis in FTIR; and Rudy Trejo in DR. One of the authors (S. Fernández-Sabido) gratefully acknowledge posdoctoral grant from *Consejo Nacional de Ciencia y Tecnología de México* (CONACYT-*México*).

REFERENCES

1. R. Gettens, *American Antiquity* **27**, 557 (1962).
2. H. Van Olphen, *Science* 154, 645-646 (1966).
3. A.O. Shepard, *Am. Antiquity* **27**, 565–566 (1962).
4. C. Reyes-Valerio, *De Bonampak al Templo Mayor. El azul maya en Mesoamérica*, Siglo XXI Ed., Mexico, 1993.
5. M. Sánchez del Río, P. Martinetto, C. Reyes-Valerio, E. Dooryhée, M. Suárez, *Archaeometry* 48 (2006) 115-130.
6. A.O. Shepard, H.B. Gettlieb, *Notes from a Ceramic Laboratory No. 1*, Carnegie Institution of Washington, Washington D.C., 1962.
7. D.E. Arnold, *American Antiquity* 36 (1971) 20–40.
8. H.E. Merwin, E.H. Morris, J. Charlot, A.A. Morris, *Publication 406*, Carnegie Institution of Washington, Washington D.C., 1931.
9. M. José-Yacamán, L. Rendón, J. Arenas, M.C.S. Puche, *Science* 273 (1996) 223–225.
10. A. Doménech, M. T. Doménech-Carbó, M. L. Vázquez de Agredos Pascual, *Archaeometry* 51, (2009) 1–20.
11. R. Kleber, L. Masschelein-Kleiner, J. Tissen, *Stud. Conservat.* 12, (1967) 41-55.

12. A. Doménech, M.T. Doménech-Carbó, M.L. Vázquez de Agredos-Pascual, *Angew. Chem. Int. Ed.* 50, (2011) 5741–5744.
13. D.G. Booth, S.E. Dann, M.T. Weller, *Dyes Pigments* 58 (2003) 73–82.
14. R. Ashok, *Artist's Pigments, A handbook of their history and characteristics*, National Gallery of Art, Oxford Unversity Press, Oxford, vol. 2, 1993, 231.
15. N. Gobeltz, A. Demortier, J.P. Lelieur, C. Duhayon, *J. Chem. Soc. Faraday Trans.* 94 (1998) 2257-2260.
16. E. Del Federico, W. Sholfberger, J. Schelvis, S. Kapetanaki, L. Tyne, A. Jerschow, *Inorg. Chem.* 45 (2006) 1270-1276.
17. M. Suárez, E. García-Romero, *Applied Clay Science* 31 (2006) 154-163.
18. F. Bosch Reig, J.V. Gimeno Adelantado, M.C.M. Moya Moreno, *Talanta* 5 (2002) 811-821.
19. C. Volzone, J. Ortiga, *Applied Clay Science* 44 (2009) 251-254.
20. C. Blanco, F. González, C. Pesquera, I. Benito, *Spectroscopy Letters* 22 (1989) 659–673.
21. A. Bakhti, Z. Derriche, A. Iddou, M. Larid, *Eur. J. Soil Sci.* 52 (2001) 683–692.
22. J. Madejová, P. Komadel, *Clays and Clay Minerals* 49 (2001) 410–432.
23. C. Dejoie, E. Dooryhee, P. Martinetto, S. Blanc, P. Bordat, R. Brown, F. Porcher, M. Sánchez del Río, P. Strobel, M. Anne, E. Van Eslande, P. Walter, *arXiv:1007.0818* (2010) (cond-mat.mtrl-sci).
24. M. Bacci, D. Magrini, M. Picollo, M. Vervat, *Journal of Cultural Heritage* 10 (2009) 275–280.
25. J. Plester, in: A. Roy (Ed.), *Artists' Pigments: A Handbook of their history and characteristics*, vol. 2, Cambridge Unversity Press, Cambridge, 1993, 37.
26. R. Corona Esquivel, M.E. Benavides Muñoz, *Boletín de Mineralogía* 16 (2005) 57.
27. C. Miliani, A. Daveri, B.G. Brunetti, A. Sgamellotti, *Chemical Physics Letters* 466 (2008) 148–151.
28. D. Ajò, U. Casellato, E. Fiorin, P.A. Vigato, *Journal of Cultural Heritage* 5 (2004) 333-348.

Mater. Res. Soc. Symp. Proc. Vol. 1374 © 2012 Materials Research Society
DOI: 10.1557/opl.2012.1383

Technical Study of a set of Metallic Artifacts from the Maya Site of Lagartero, Chiapas, Mexico

Gabriela Peñuelas Guerrero[1], Ingrid Jiménez Cosme[1], Pilar Tapia López[1], José Luis Ruvalcaba Sil[2], Jesús Arenas[2], Aurore Lemoine[2], Jannen Contreras Vargas[1], Patricia Ruiz Portilla[1], Sonia Rivero Torres[3].
[1]Escuela Nacional de Conservación, Restauración y Museografía, ENCRyM, Instituto Nacional de Antropología e Historia, INAH. General Anaya No. 187, Col. San Diego Churubusco México DF 04120, México. e-mail: gapenuelas@gmail.com
[2]Instituto de Física, Universidad Nacional Autónoma de México UNAM. Apdo. postal 20-364, México DF 01000, México. e-mail: sil@fisica.una.mx.
[3] Dirección de Estudios Arqueológicos, Instituto Nacional de Antropología e Historia (INAH). Lic. Verdad No. 3, Centro Histórico, Mexico DF 06060, Mexico.

ABSTRACT

During the excavations made at a burial of the post-Classic Mayan period (1220-1521 A.D.) in the pyramid number 2 of the pre-Hispanic site of Lagartero, Chiapas, Mexico, a set of four small metallic artifacts depicting reptile's heads, were recovered. The objects were in poor conservation conditions and were taken to the Metal Conservation Laboratory of the National School of Conservation (ENCRyM- INAH) for suitable cleaning and conservation treatments.

Analyses allowed identifying important technological features such as gilding remains. The analytical techniques included optical microscopy followed by X-Ray Fluorescence Spectroscopy (XRF), and Scanning Electron Microscope-coupled with Energy Dispersive Spectrometer (MEB-EDS). For the elemental depth profile a combined Particle X-Ray Emission Spectroscopy and Rutherford Backscattering Spectrometry (PIXE–RBS) analysis was carried out.

The results indicate that the objects are made of a copper alloy, and then gilded probably using electrochemical replacement gilding (thickness has less than 1 μm). This technology has been observed in other metallic objects recovered from the Chichen-Itza Cenote in Yucatan, but it was not available in Mesoamerican areas so far. So, it could imply that these artifacts might have been imported from South American areas.

INTRODUCTION

The set of metallic artifacts was recovered in 2009 at the 12th season of excavations at the Lagartero Project. The items were found associated to several bones in a multiple burial, among 15 skulls were found nevertheless any complete body was recovered. In addition to these artifacts, a complete bell and a textile fragment associated to it were also discovered.

The Mayan site of Lagartero is located amongst the Colon Lake, nearby the Lagartero and Grijalva rivers at the border of Mexico and Guatemala in Chiapas (Figure 1). This archaeological site is formed by a series of small islands; at the most important one, called *El Limonar*, the multiple burial was discovered inside the pyramid number 2, the tallest structure of the site.

Although the archaeological investigations are still in course, it is known that this Mayan site was occupied since the pre-Classic until the post-Classic period (100 A.D to 1200-1521A.D). The location and the pottery found in the site, allowed the archaeologists to conclude that Lagartero had commercial relationship with many foreign countries in Mesoamerica and, maybe even, in South America [1].

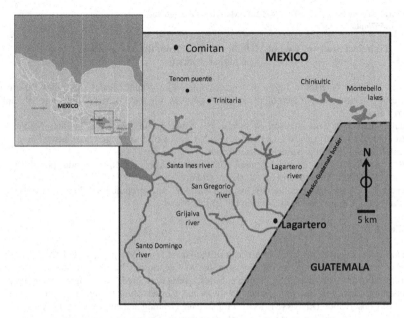

Figure 1. Location of the Maya Site of Lagartero, Chiapas, Mexico.

The artifacts were made by lost-wax casting within an unusual shape; at the center each one had four little square-shaped knobs that remember a reptile head. Judging from the dimensions (shown in Figure 2) they probably were used as collar beads or knitted to a textile. Either hypothesis is supported by the fact that each item has a pair of minute perforations at each side. In addition, one of the artifacts had vestiges of cotton fibers in one of these perforations. The artifacts were taken to the Metal Conservation Laboratory of the National School of Conservation (ENCRyM-INAH) for suitable conservation treatment.

Recently the Metal Conservation Laboratory adopted the methodology developed by the ANDREAH Project (Non Destructive Analyses for Art, Archaeology and History Studies) to characterize metallic collections [2]. This article describes the results of the technical examination of the four metal artifacts carried out by the joint work of professionals of the Metal Conservation Laboratory of the ENCRyM and the Instituto de Fisica of UNAM, in order to identify their technological characteristics to apply suitable conservation treatments and to help to its symbolic understanding and give insights about their probable provenance.

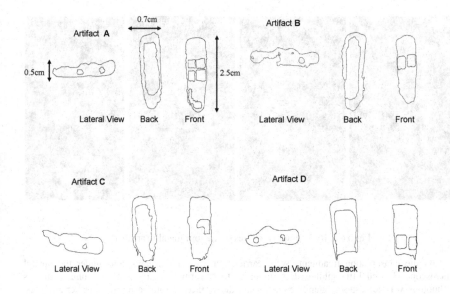

Figure 2. The Lagartero artifacts. Drawings by P. Ruiz Portilla.

EXPERIMENTAL METHODOLOGY

Frequently, during the conservation treatments is possible to identify important technological features inaccessible by other means, so the findings of the professional conservator constitute a sound source of knowledge about the object for other fields, this case is an example. Therefore the examination made for conservation purposes and even the conservation treatments are included in this section.

The four objects were recovered in poor conservation conditions due to corrosion caused by the chemical interaction of the copper alloy and the burial context [3]. Three of the items were incomplete; the metal core was lost due to corrosion, and all of them presented a mixture of active and passive corrosion products and soil deposits on the surface.

It was followed a detailed examination of the objects using optical microscope before and after the conservation treatment. In general terms, during the first examination we were able to locate the fiber textiles and the gilding remainders beneath the corrosion layer. Additionally, radiographic plates were taken to each artifact in order to identify their morphology and to assess their conservation condition (Figure 3). As the following figure shows, the objects have a perfectly defined quadrangular relief design and two lateral small holes at each side.

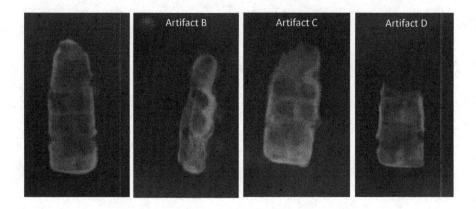

Figure 3. Radiographic images of the four metallic artifacts.

As the conservation treatments were carried out, the first examinations under the optical microscope showed little gilding dots. In order to confirm the gilding and to understand its technology was necessary to stabilize chemically the surfaces before performing further analyses. An elemental composition analysis was undertaken before conservation treatments, using a portable X-Ray Fluorescence Spectrometer of the Physics Institute (UNAM), with Mo X-ray tube with voltage of 45kV, a current of 0.2mA, and a Si-PIN detector [5]. To understand the gilding technology the objects were also analyzed using a Scanning Electron Microscope-coupled with Energy Dispersive Spectrometer –MEB-EDS- JEOL JSM 6450, in the low vacuum mode, at the MEB laboratory of the INAH.

Because of their conservation state the conservation treatments objective was to recover the artifacts' shape and legibility; consequently, the conservation treatments were focused on removing the disfiguring corrosion products, and the soil concretions.

The surface of the artifacts was mechanically cleaned with a pointless wooden stick, soft brushes and in necessary cases with a scalpel under low magnification (5x), aided by rising with distilled water. Nevertheless a local chemical cleaning was required to eliminate the undesired corrosion products using a chelating agent: disodic EDTS in 5% (w/v) aqueous solution, pH 5, and rinsed with acetone in order to eliminate the chelating remains. Finally, the surface was stabilized by using a benzotriazole (BTA) 1% (w/v) solution in ethanol. This material was chosen since it forms a protective film of Cu(I) benzotriazole complex that retard the cathodic reduction of oxygen [4]. In order to continue with the characterization it was unnecessary to place any protective layer on the surface of the items (Figure 4).

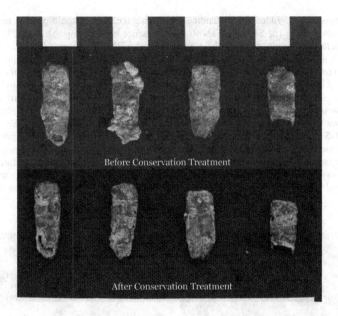

Figure 4. View of the four metallic artifacts before and after conservation treatment.

Once chemically stabilized, the artifact A was chosen because it had most amount of gilding vestiges. It was analyzed again by using a JEOL JSM 5600LV SEM-EDS, in low vacuum mode (8Pa), at the Microscope Laboratory of the Physics Institute- UNAM.

In order to characterize the composition of the gilding layer the beads were analyzed with Particle Induced X-ray Emission technique (PIXE) using a 3 MeV proton external beam with 1.5 mm diameter beam spot. Various regions of the artifact A were analyzed: with and without gilding. Also, in order to establish the thickness of the gold layer and identify the elemental profile at each layer Rutherford Backscattering spectroscopy (RBS) was applied simultaneously to proton external beam PIXE [6-9]. But since the depth resolution of the protons beam that can reach a depth of analysis of 15μm [8-9] in this kind of matrix is too low to distinguish the thin gilding layer, it was necessary to switch to a 3 MeV He beam in vacuum with a 1.5 mm diameter beam spot. In this way the beam may reach an analysis depth of only a 3 μm with a significantly increase in the depth resolution. Using this method it was possible to measure the elemental depth profile in order to verify the plating composition [10, 11].

RESULTS AND DISCUSSION

After the gilding remains discovered during the first examination, the *in situ* XRF analysis revealed the presence of copper, gold and small amounts silver. These results could be due to two possibilities: the presence of a ternary alloy, common in the Mesoamerican region, or by a gilded copper alloy.

If the objects were gilded, understanding the gilding technology could provide information about their provenance. The SEM-EDS showed the presence of C, O, Mg, Al, Si, Fe, P, S, Cl, K, Ca, at the surface composition, all of them characteristic of the burial context and the corrosion layer; besides these elements, the data confirm the presence of copper and interestingly gold and silver only in one region. These results allowed considering the gilding option almost for sure.

Nor the XRF or the EDS results showed arsenic, lead or tin in the copper alloy, all these elements are common constituents of the known composition of Mesoamerican artifacts alloys [12,13]. Consequently this fact can be revealing about the provenance of the analyzed objects.

Once removing the disfiguring corrosion and soil deposits, the artifact A was analyzed again. The SEM-EDS results showed the differences between the copper alloy matrix, the gilding layer and the corrosion products beneath and over the plating (Figure 5). Also the 1,300x magnification indicate a linear pattern probably due to the polish of the plating during the manufacture process, besides the evident presence of multiple pores which could help to identify the manufacture technique, and a few cracks most likely produced by the tensions of corrosion and the burial conditions.

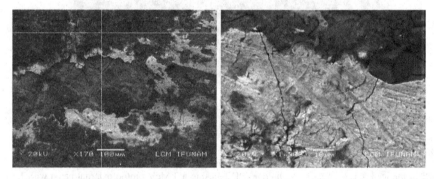

Figure 5. SEM images of the Artifact A surface. Left: 100x magnification of the gilded and corroded zones. Right: 1300x magnification of the gilding showing linear pattern probably due to polishing, pores and cracks.

According to these results, it can be suggested that the poor conservation condition of the artifacts was enhanced by the morphology of the gilded layer, due to the presence of pores that allowed the gold-silver alloy to interact with a less noble and more active copper matrix, in a galvanic corrosion phenomena also promoted by the presence of humidity and corroding ions in the burial soil. As the copper began to corrode in the interface between the gilding and the cuprous bulk, the volume increase of corrosion products caused the detaching of the gilding layer until almost losing it all.

On other hand, the PIXE spectra (Figure 6) confirmed the presence of the gold-silver alloy even on the corroded areas.

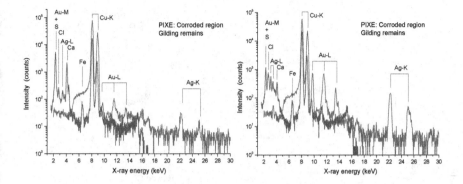

Figure 6. PIXE spectra of two different regions of the artifact A indicating the main surface constituents: Cu, Ag and Au.

As a result of the advanced deterioration of the artifacts, the remains of gold and silver alloy were not as evident as the copper in the previous analysis.

RBS spectra from three different regions of the artifact A are shown in figure 7. The corresponding elemental depth profile measured by RBS is shown in Figure 8. Interestingly the presence of gold and silver vestiges in all the regions at the surface only, with gilding vestiges and without them, remains constant, obviously showing different amounts. However from the RBS signals, the Au-Ag proportion of the gilding alloy conserve its ratio in all the regions. Moreover, through RBS a silver diffusion zone as an interface between the copper alloy matrix and the gilding was detected. The corrosion linked to O contents is higher also at the inter-phase between the plating and the Cu support. Also by means of RBS it was possible to measure the thickness of the gilding, finding that was about 0.4 μm. These results point out to the use of electrochemical replace gilding, since it is the most likely to achieve a continuous plating layer that could range usually between 0.5 to 2 mm [10-11].

As described by Lechtman, the electrochemical replacement gilding requires immersing the metallic objects, often cuprous, in a hot bath solution within ionized gold and silver. The gilding is achieved by electrochemical reactions between the objects' metallic surface and the ions in the solution; as it was acidic the metalsmiths had to neutralize it and then heat to enhance the bonding by diffusion between the gold-silver alloy and the copper alloy matrix [14].

Even though Fray Bernardino de Sahagún described during the 16th Century in the Florentine Codex, the lost-wax casting and the depletion gilding manufacture techniques [15], nowhere established the electrochemical replacement as a method employed by Mesoamerican metalsmiths [16].

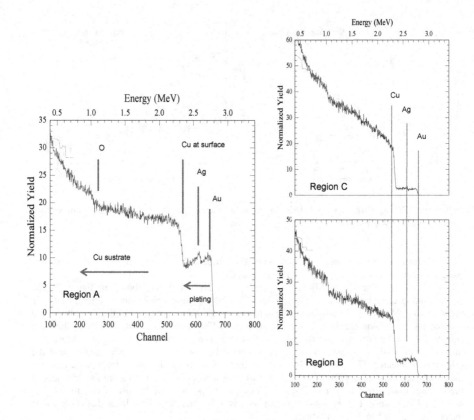

Figure 7. RBS spectra of different regions of the artifact A. Left: Spectrum of the gilding remains zones showing higher amounts of Au and Ag at the surface. Right: Two spectra of zones without evident gilding but showing that the Au-Ag ratios at the surface remain constant.

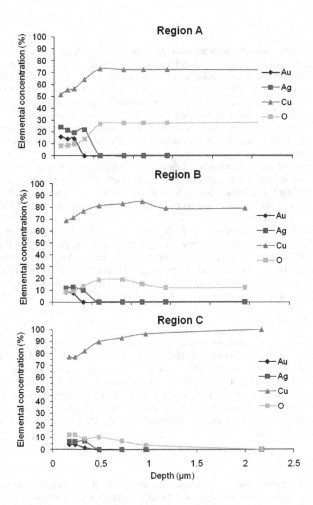

Figure 8. Elemental depth profile of the three regions (A, B and C) of the RBS spectra shown in figure 7. The gilding thickness may reach 0.4 μm while a silver diffusion is observed in all the cases.

CONCLUSIONS

Interdisciplinary work provides different kinds of knowledge about the objects; in this case the results of the different examinations are useful for conservation, technological and provenance aims. This research provided valuable information to understand the manufacture and gilding techniques, according to the SEM-EDS and PIXE-RBS results it could be said that the Lagartero artifacts were gilded by the electrochemical replacement method using a gold-silver solution.

From these results it can be said that the current deterioration was due to a galvanic corrosion phenomenon caused by the interaction of the two different oxidation potential alloys in the artifacts, the presence of pores and the reduced thickness of the gilded layer acting synergically with the burial environment.

As previously established, the technological data about gilded items are very rare in Mexican archaeometallurgy. One of the few examples is a pendant found at Caxonos, Oaxaca, that was gilded by depletion gilding [9], a method described by the etnohistoric sources and totally different to what was found at these four artifacts and on previously analyzed artifacts of the Chichen-Itza Cenote [10-11, 17].

Although there is still few data to support further conclusions, the available information and the absence of arsenic, lead or tin in the constituent copper alloys allow to propose that the objects may were not made in Mesoamerica [13]. The electrochemical replacement method was developed at the Peruvian zone [14], reason why it could be suggested that these artifacts arrived to the Mayan region by means of commercial exchange with other regions, in the same way as it has been suggested for various objects recovered from the Chichen Itza [10-11, 17].

ACKNOWLEDGEMENTS

The authors would like to thank to Karim López and Francisco Jaimes for the Pelletron Accelerator operation as well to Mario Monroy and Jaqueline Cañetas for the SEM-EDS analyses done at the IF-UNAM; Gerardo Villa for SEM- EDS at the Subdirección de Laboratorios INAH; and Dr. Josefina Bautista and Mario Reyes Valdez for the radiographies taken at ENCRyM-INAH.

Authors acknowledge the supports of UNAM PAPIIT IN403210 project as well as the CONACYT Mexico grant 131944 to carry out this study. This research has been performed in the frame of the Non Destructive Study of the Mexican Cultural Heritage ANDREAH network (www.fisica.unam.mx/andreah).

REFERENCES

1. E. Rivero. S. Torres. *Informe de la 12° Temporada de Campo del Proyecto Arqueológico de Lagartero, Municipio, La Trinitaria, Chiapas*, México, Dirección de estudios Arqueológicos del Instituto Nacional de antropología e Historia, INAH, 2010. unpublished.
2. G. Peñuelas Guerrero. *Caracterización por medio de análisis instrumentales de los materiales constitutivos de la orfebrería de la tumba 7 de Monte Albán, Oaxaca*, Dissertation Thesis on Conservation, Escuela Nacional de Conservación, Restauración y Museografía- Instituto Nacional de Antropología e Historia. México, 2008.
3. L. Selwyn. *Metals and corrosion. A handbook for the conservation professional*. Canadian Conservation Institute, Ottawa, 2004, 23.

4. D. Scott. *Copper and bronze in art. Corrosion, colorants, conservation,* Getty Conservation Institute, Los Angeles, 2002, 377.

5. Ruvalcaba Sil, J.L., Ramírez Miranda, D., Aguilar Melo, V. and F. Picazo, *X-ray Spectrometry* 39 (2010) 338-345.

6. J.L. Ruvalcaba Sil. Las técnicas de origen nuclear: PIXE y RBS, in: *La Ciencia y el Arte,* vol I. M.A. Egido y T. Calderón (coords.), Instituto del Patrimonio Histórico Español. IPHE-CSIC, Madrid, 2008, 151-172.

7. G. Demortier, J.L. Ruvalcaba-Sil. Non-destructive ion beam techniques for the depth profiling of elements in Amerindian gold jewellery artifacts in: *Cultural Heritage Conservation and Environmental Impact Assessment by Non-destructive Testing and Microanalysis,* R. Van Grieken & K. Janssens (eds.), A.A. Balkema Publishers, London, 2005, 91-100.

8. G. Demortier, J.L. Ruvalcaba, *Nuclear Instruments and Methods in Physics Research B* 239 (2005) 1-15.

9. E. Ortiz Díaz, J.L. Ruvalcaba Sil. An historical approach to a gold pendant: the study of different metallurgic techniques in ancient Oaxaca, Mexico, during the late postclassic period, in: *Archaeometallurgy in Europe: 2nd International Conference, Aquilea, Italy, 17-21 June 2007: selected papers,* P. Craddock, A. Giumlia-Mair, A: Hauptman (eds.), Associazione Italiana di Metallurgia, Milan, 2009, 511-518.

10. J. Contreras, J.L. Ruvalcaba, J Arenas. *Non destructive study of the gilding of tumbaga artefacts from Chichen-Itza cenote,* Metal 07, ICOM-CC Triennial conference, C. Degrigny ed., Amsterdam, 2007, 53-56.

11. J. Contreras, J.L. Ruvalcaba Sil, J. Arenas Alatorre. *Non destructive study of gilded copper artifacts from the Chichen-Itza Cenote,* in Proceedings of Particle Induced X-rays Emission and its Analytical Applications, PIXE 2007, Puebla, Mexico. UNAM. J. Miranda & J.L. Ruvalcaba (eds.), 2007 (electronic publication in CD).

12. J.L Ruvalcaba Sil, G. Peñuelas Guerrero, J. Contreras Vargas, E. Ortiz Díaz, E. Hernández, *ArcheoSciences Revue d'archéométrie* 33 2009, 289-297.

13. N. Schulze. "For Whom the bell tolls" Mexican copper bells from the Templo Mayor Offerings: Analysis of the production process and its cultural context, *Material Research Society Symposium Proceedings,* Vandiner, P.B., Mc.Carthy, B., Tykot R., Ruvalcaba J.L. & F. Casado (eds.), Vol. 1047, Materials Research Society, Warrendale, 2008, 195-204.

14. H. Lechtman. *Journal of Metals* 31 (1979) 154-160.

15. B. Sahagún de. *Historia General de las Cosas de la Nueva España,* Porrúa Ed., Mexico, 2006. 503-505.

16. W. Bray. Techniques of gilding and surface enrichment in pre-Hispanic American metallurgy, in: *Metal plating and patination: cultural, technical and historical developments,* S. LaNiece & P. Craddock (eds.), Butterworth Heinemann, Oxford, 1993, 182-192.

17. J. Arenas Alatorre, J. Contreras Vargas, J.L. Ruvalcaba Sil. Microestructural Study of Gilded copper artifacts from the Chichén-Itzá Cenote, in: *2nd Latin-American Symposium on Physical and chemical methods in Archaeology, Art and Cultural Heritage Conservation & Archaeological and Arts Issues in Materials Science – IMRC 2009, México: selected papers,* J.L.Ruvalcaba, J.Reyes, J. Arenas & A.Velázquez (eds.), Universidad Nacional Autónoma de México- Universidad Autonoma de Campeche, Instituto Nacional de Antropologia e Historia, Mexico, 2009. 67-71.

Mater. Res. Soc. Symp. Proc. Vol. 1374 © 2012 Materials Research Society
DOI: 10.1557/opl.2012.1384

Identifying the Criteria for Determining Status Burials by Results of Integrated Analysis

Irina A. Saprykina[1] and Olga V. Zelentsova[1]
[1] Institute of Archaeology Russian Academy of Sciences,
19, Dm.Ulyanova str., Moscow, Russia, 117036. e-mail: dolmen200@mail.ru

ABSTRACT

Chemical-technology investigation was carried out for 95 belt sets from the burials of 8th–11th centuries. In its course three groups of burials were separated on the basis of archaeology-statistical features; the Group 3 was identified as status one. All the groups contain belt sets made from the cooper-based alloys which include silver (6.98−92.87%) and amalgam. The major scheme of the belt sets (Groups 1, 2, 3) production is casting in moulds with print of used plaques; there are belt sets in Groups 1 and 2 which were made in different way (lost wax casting, casting in moulds with print of special matrix, forging). Chemical and technology parameters of the belt sets from the elite burials of the Group 3 are similar to the parameters of the belt sets discovered in the common burials of the Groups 1 and 2. There are a number of burials in the Group 1 the belt sets of which according to their chemical-technology characteristics could be attributing to the "award".

INTRODUCTION

Chemical and technology investigation of the belt sets items is a part of investigation dedicated to the reconstruction of ancient society structure. In particular this investigation is aimed at separation of true criteria for the identification of status burials entering into the composition of the burial grounds and its historical interpretation. In historiography the belt sets, particularly its series and metal type [1-4], are considered as one of the indicators of high social status of the buried.

This investigation is based on materials from the Krjukovsko-Kuznovsky burial ground situated in Central Russia (river Tsna − the basin of river Volga). In 1930-th this monument was investigated by one of the first Soviet archaeologists, antiquarian and founder of Morshansk museum (Tambov region) P.P. Ivanov, who excavated 586 burials dating to the 8th–11th cc. [5]. In the course of investigations it was established that burial ground had belonged to Mordva − one of the povolzhsko-finnish tribes inhabited Povolzh`e from the first centuries AD.

In the basin of river Tsna 15 burial grounds left by Mordva-tribe and dated to the 8th–11th cc. are known, of which the Krjukovsko-Kuznovsky is one of the most completely investigated. It is a common cemetery with burials made both by cremation (14%) and inhumation (86%). Another key element of the burial ceremony is characteristic grave goods presented by the tools, weapon, everyday objects and also by the huge amount of metal ornaments especially belt sets intended for the suit's decoration.

The Krjukovsko-Kuznovsky burial ground is known by the bad safety of anthological dates that is why in the course of investigation we had to deal with the results of archaeological analysis of the grave goods and statistic data. Distribution of the burials on the basis of sex and age showed that the third part of graves is male. The belt sets with metal plagues are fixed in 140 graves that make up 76% from the whole number of male graves. Burials the belt sets of which are presented by one item aren't considered in further investigations (45 burials). Analytical selection is reduced to 95 male graves with full complete belt sets was discovered.

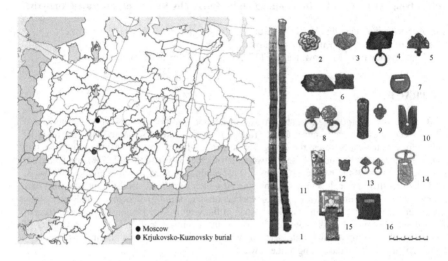

Figure 1. Examples of belt sets from Group 1: *1* – gr.505, *2* – gr.192, *3* – gr.201, *4* – gr.312, *5* – gr.409; Group 2: *6* – gr.196, *7* – gr.424, *8* – gr.511, *9* – 451, *10* – gr.348; Group 3: *11* – gr.407, *12* – gr.153, *13* – gr.175, *14* – gr.69, *15* – gr.206, *16* – gr.165.

Statistical analysis of the obtained material gave us reason to separate a few groups of burials differed in quantity and composition of grave goods and, in particularly, in presence of weapon. The first and the most numerous (64%) group of burials (Group 1) is characterized by the belt sets consist of 25–40 items and presence of one or two types of weapon in the grave (axe and spear or axe and arrows). The second group (30%; Group 2) is characterized by the belt set with plaques which number changes from 40 to 120 and presence of three and more types of weapon: an axe, spear and quiver with arrows (5–20).

The third rare group of burials (6%; Group 3) consists of 4 and more types of weapon including such rare types as ice pick and mattock and also of imports and exclusive weapon usually discovered in the military elite graves [6] such as iron stiletto in scabbard and blades. The belt sets from Group 3 consist of 25–40 items that make this group similar to the Group 1. The graves of Group 3 are the original centers around which another graves were concentrated. The burial ceremony of this group is specified by the large burial peats which are exceeded average anthropometric parameters, first of all in depth (it reaches 1.5 m in depth). In the graves remains of burial food and ceremonies connected with fire are found.

Consequently the group of status graves was separated on the basis of archaeological investigation and statistic data whose part in the general amount of graves of the Krjukovsko-Kuznovsky burial ground (586 graves) doesn't exceed 2% (13 graves).

Let us consider in what way obtained information correlates with results of the chemical analysis.

EXPERIMENTAL DETAILS

Chemical-technology analysis consisted of: 1) analysis of chemical composition of the base metal of the belt sets items with help of XRF [7]. The measurement procedure is standardized, measurement accuracy is set for each individual element; 2) the reconstruction producing technologies of objects were examined with help of binocular microscope Motic BA-300, photofixing was carried out using digital camera Moticam 2300 (compulsory increase 0.5x).

The general investigation base consists of 210 samples obtained from the belt sets items from 26 graves. The data obtained for the belts discovered in women's graves has been excluded from general investigation. Consequently, the analytic selection from male burials consists of 174 samples from 20 burials.

Examination of the traces of technological operations has been conducted for all belt sets that arranged the analyzable selection (more than 800 items). Detailed description of the results of this examination is given for the belt sets for which chemical composition of metal is known; description of the results of technological examination of other belts is given in total.

Unfortunately, there was no possibility for conducting a number of special studies such as metallography, SEM and so on (the works were carried out in museum closed for reconstruction), that is why the problem of using special technique as, for example, silvering [8] is still unsettled.

DISCUSSION

The belt sets from Group 1 were made of different types of cooper-based alloys: tin-lead bronze, double and triple brasses, multi-component alloys (Table 1). Metal of the belt sets from Group 1 is specified by the variability of presence of such alloying parts as tin (from 10.95 to 26.07%; for example, grave 406) and lead (from 12.03 to 27.86%; grave 267). Metal of the three studied belt sets contained silver as one of the components (6.98−45.17%); one belt set was made from low-grade silver (40.82−68.91%; grave 192). Items of three belt sets produced from silver-based alloys (gr.192, 201, 505) were gilded in amalgam technique.

The most part of items of the belt sets from Group 1 were made by casting; the major schemes are presented by casting in two-part moulds with print of used plaques that in its turn was made by casting in moulds with print of special matrix or by lost wax casting (multiple copying). In the course of such method typical defects that were transferred from finished items employed as matrixes to the castings are usually marked: first of all it is asymmetry, holes on the surface of the finished items (technical holes, cavities) and diffused surface.

The package of the belt sets not only by plaques made from different types of alloys (Table 1, graves 201, 409 and so on) but also by plaques made by different kinds of casting could be treated as common practice: for example, in the belt set originated from grave 201 elements made by lost wax casting and by casting in two-part moulds with print of used plaques are combined. There are also belt sets combine "title" plaques alongside with the plaques made by print of this "title" plaques (grave 312); its finale package could be produced either at once or in several stages.

Table 1. XRF data for the investigated samples from Group 1.

References	Amount of samples	Type of Alloys	Min\Max	AuHg
Grave 192	18	CuZn+Ag CuSnZn+Ag CuSnPbZn+Ag	Ag: 40.82-68.91%	+
Grave 201	6	CuAg CuPbZn+Ag	Ag: 30.64% Ag: 29.36-45.17%	+
Grave 229	3	CuSnPbZn CuSnPb		
Grave 267	11	CuSnPb		
Grave 312	2	CuSnPbZn		
Grave 361	3	CuSnPb CuSnPbZn		
Grave 406	11	CuSnPb		
Grave 409	9	CuSnPbZn CuSnPb+Ag CuSnPbZn+Ag	Ag: 22.7-39.54% Ag: 6.98-18.88%	
Grave 505*	22	CuZn+Ag	Ag: 15.97-36.01%	+

*Results of chemical-technology investigation of the belt set from the grave 505 were partly published [10].

General technological scheme, unification of the producing processes marked the majority of studied belt sets (80% of the selection), frequency of occurrence the belts from Group 1 on a wide area – all the evidences goes to prove that the belts production was massive.

At the same time Group 1 consists of the belt sets (graves 192, 201 and 505) that were made by special casting in moulds with print of individual matrixes of different types; traces of the multilevel technological operations left on the surfaces of the belt's items are clearly visible. Beyond that point on the belt from grave 201 the traces of the independently executed repair were noticed: cover of the buckle was replaced. It was nailed to the own frame by the iron pin and after it fastened for the additional safety by the leather strap. In all likelihood in the massif of the burials from Group 1 the circle of so called military (warriors) graves that are marked by the presence of specially manufactured belt sets regarded as an "award" might be identified.

Analytical selection of the belt sets from the graves of Group 2 shows that these belts were made from common set of cooper-based alloys (CuSn, CuSnPb, CuSnPbZn); however unlike the belts from Group 1 these alloys contain more tin and lead (Table 2). The traces of gilding that was discovered on the surface of the belt produced from lead brass (grave 196) could be treated as an unusual fact. Separate plaques of the belt set from grave 472 were produced from silver-based alloy obtained in the result of the dirham's remelting (bismuth 0.67−1.72%) [9].

The vast majority of belt sets from Group 2 is made by casting in two-part moulds with print of used plaques. Details obtained in the course of such technological process have numerous defects transferred from the used "matrixes": asymmetry, technical holes and diffused surface. A considerable part of the belt sets was made by more difficult scheme – lost wax casting when lost wax model was obtained by casting in special matrix (belts of "Saltovo" horizon, graves 196, 511); casting in the no-clay moulds was also known (grave 424). Singular belts were made by forging (laminate belt set, grave 348).

Table 2. XRF dates for the investigated samples from Group 2.

References	Amount of samples	Type of Alloys	Min\Max	AuHg
Grave 139	6	CuSn CuSnPbZn CuSnPb	Sn: 33.8% Pb: 33.11-43.62%	
Grave 196	1	CuPbZn		?
Grave 228	3	CuSnPb CuSnPbZn		
Grave 348	5	Ag+CuPb	Ag: 78.99-92.87%	
Grave 424	2	CuSnPb		
Grave 451	9	CuSn CuSnPb	Sn: 41.32%; Sn: 19.55-37.97% Pb: 1.38-44.51%	
Grave 472	10	CuPbZn+Ag CuPb+Ag CuSnPb+Ag	Ag: 14.78-58.24%	
Grave 511	6	Ag+CuPb	Ag: 60.19-90.1%	?

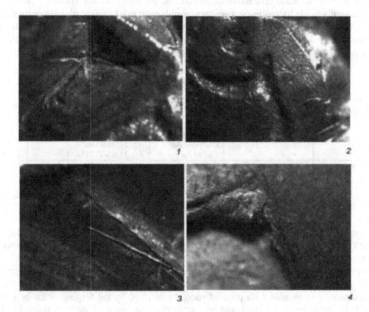

Figure 2. Examples of some defects on finished items: cavities, cutting lines, layering of amalgam (10x, 0.5x).

The standard of manufacture, massive production is typified by the points noticed in the course of the examination of the whole discovered items; fasteners are usually treated as such points. For the rare exceptions pins made of cutted rods of common size (d=2 mm, h=10-15 mm) were used as fasteners. As a rule such pins were put into the special holes made in the moulds and casting was produced by the "refill" method. From our point of view the plaques with typical cavities left by the pins knocked out of the casting in the course of its beat out of the mould testify for the belt's producing as a massive one. Such plaques were "decorated" by the holes pierced through ornament's relief (sometimes through the gilding surface) and were attached to the belt by pins with unrivet heads.

Rare type of belt set from Group 2 is presented by the belt from grave 228 in the composition of which there is niello tip. Belt's details are produced from the cooper-based alloys (Table 2); the tip is made by casting in two-part moulds with print of used tip. The casting is massive; its relief is additionally cut and deepened; inside relief there is black-colored paste of glassy structure.

Burials of the Group 3 are status ones; it is typified by the presence of 4 and more types of weapon including rare one and imports. There are only 13 such burials in the whole selection. Chemical composition of metal of the belts from the graves 69, 206, 407 was studied (Table 3). These belt sets have the same characteristic features as the sets from the Groups 1 and 2. The belts were generally made by casting in two-part moulds with print of used plaques or by the lost wax casting when lost wax model was obtained by casting in special matrix. Separate items of the belts from Group 3 have the same defects as the major part of the belts from Groups 1 and 2. Some belts could be arranged of items produced in several stages, not by special order.

Table 3. XRF dates for the investigated samples from Group 3.

References	Amount of samples	Type of Alloys	Min\Max	AuHg
Grave 69	7	CuPbZn+Ag	Ag: 21.53-43.17%	
Grave 206	29	CuAg	Ag: 22.93-67.26%	+
		CuPb+Ag	Ag: 42.15-72.13%	
		CuPbZn+Ag	Ag: 51.02-75.02%	
		CuZn+Ag	Ag: 35.02-37.71%	
Grave 407	11	CuPbZn+Ag	Ag: 61.12-80.5%	

CONCLUSIONS

Therefore, 70% from the studied belt sets were made by the casting in two-part moulds with print of used items. According to the obtained information such belt sets were arranged from the items produced in different times; consequently the belt was arranged from the certain items as they were revealed. Such sets were marked in all three groups; in Group 3 they are prevalent.

30% of the belt sets were made by the lost wax casting by melting mould, casting in the no-clay moulds, casting in moulds with print of special matrix or by the forging. Such belts were usually made by the order of a certain customer or in the first-rate workshop accommodated the certain class of society. These belts are illustrative of the selection from Group 1 and 2.

Studied belt sets were made from the cooper-based (tin-lead bronze, double and triple brasses, multi-component alloys) and silver-based alloys. The alloys with high percentage of silver are rare in the selection (see Table 1-3); they belong to the Group 2.

Presence of individual belt sets in the first two groups probably indicates the role of a belt as a special honorary item that were granted to the distinguished members of society (militaries or warriors). At the same time according to the results of the chemical-technology analyzes of the belts from the third elite group of burials individual belt didn't consider as required part of the elite suit unlike the blades, flails, festive ware that underlined the status of its owner. From one side it points to the certain differentiation characterized the studied society, from the other it is said about mobility of its members.

ACKNOWLEDGMENTS

This research was produced with support of RFFI grant 08-06-00299a.

REFERENCES

1. G. Laszlo. *Etudes archeologiques sur l'histore de la societe des Avares, FH.V.XXXIV.* Budapest, 1955, 176.
2. S. A. Pletneva. *From nomad to cities. Saltovo-Mayazkaya culture.* Moscow, 1976, 164 (in rus.).
3. G. E. Afanasev. *Don Alans. Social structure of alan-asso-burtas population of Middle Don river basin.* Moscow, 1993, 45 (in rus.).
4. M. Shulze-Dorrlam. *Byzantinische Curtelschnallen und Gurtelbeschlage im Romisch-Germaniscen Zentralmuseum,* Meinz, 2009, 308.
5. Materials of Mordovian History of 8th - 11th cc: *Journal of archaeological excavation of P.P.Ivanov.* Morshansk, 1952 (in rus.).
6. R. Laszlo. *A Korosi honfoglalas Kori femetok. Regeszeit adatok a Felso-Videk X. szazadi forfenetehez.* Misskols, 1996.
7. N. Eniosova, R. Mitoyan, T. Saracheva. "Methods for the study of nonferrous chemical compounds" in *Nonferrous and precious metal alloys in Eastern Europe in the Middle Ages.* Moscow, 2008 (in rus.).
8. A. Giumlia-Mair. Colouring treatments on ancient copper-alloys, *La Revue de Métallurgie-CIT. Science et Génie des Matériaux,* 2001, 767-774.
9. N. Eniosova, R. Mitoyan. Arabic Coins as a Silver Source for Slavonic and Scandinavian Jewelers in the Tenth Century AD, *Proceedings of the 37th International Symposium on Archaeometry,* I. Turbanti-Memmi ed., XLV, Springer, Heidelberg, 2011, 583.
10. I. Saprikyna, O. Zelentsova, R. Mitoyan. Warrior's Belt from the Middle Volga Burial Ground X a.d. – Technology and Origin, in: *2nd Latin-American Symposium on Physical and Chemical Methods in Archaeology, Art and Cultural Heritage Conservation. Symposium on Archaeological and Arts Issues in Material Science. IMRD 2009. Selected Papers.* Edited by J.L. Ruvacalba, J.Reyes, J. Arenas, A. Velazquez. Universidad Nacional Autonoma de México, Universidad autónoma de Campeche, Insituto Nacional de Antropologia e Historia, Mexico, 2010, 62-66.

Mater. Res. Soc. Symp. Proc. Vol. 1374 © 2012 Materials Research Society
DOI: 10.1557/opl.2012.1385

New Insights into Ancient Maya Building Materials: Characterization of Mortar, Plaster, and Coquina Flagstones from Toniná

Francisco Riquelme [1, 2], Martha Cuevas-García[4], Jesús Alvarado-Ortega[4], Shannon Taylor[1,5], José Luis Ruvalcaba-Sil[2], Carlos Linares-López[6], Manuel Aguilar-Franco[2], Juan Yadeun-Angulo[7]

[1] Posgrado en Ciencias Biológicas, Universidad Nacional Autónoma de México, UNAM. Circuito Interior s/n, Ciudad Universitaria, Mexico DF 04510, Mexico.
e-mail: riquelme.fc@gmail.com
[2] Instituto de Física, Universidad Nacional Autónoma de México, UNAM. AP 20-364, Mexico D.F. 01000, Mexico. e-mail: sil@fisica.unam.mx
[3] Dirección de Registro Público de Monumentos y Zonas Arqueológicas. Instituto Nacional de Antropología e Historia. Av. Victoria 110, Copilco El Bajo, Mexico DF 04510, D.F, Mexico.
[4] Instituto de Geología, Universidad Nacional Autónoma de México, UNAM. Circuito de la Investigación Científica s/n, Ciudad Universitaria, Mexico DF 04510, Mexico.
[5] Center for Materials Research in Archaeology and Ethnology. Massachusetts Institute of Technology, Cambridge, MA 02139-4307. USA.
[6] Instituto de Geofísica, Universidad Nacional Autónoma de México, UNAM, AP 120-364, Mexico D.F. 01000, Mexico
[7] Dirección de Estudios Arqueológicos. Instituto Nacional de Antropología e Historia. INAH. Calle Primo de Verdad 3, Centro Histórico, Mexico DF 06050, Mexico.

ABSTRACT

This work shows current research on lithic raw material used by the ancient Maya of Toniná. The core of the city of Toniná lies on a steep-sided hill of calcareous sandstones from the shallow marine deposits dated as Oligocene, in the Chiapas Highlands of Southern Mexico. Results of paleontological fieldwork in Toniná show several biostrome sediments mound-like with tabular bafflestones and large coquina flagstones, which are sheet-like rocks enriched with fossil mollusk shells, corals, encrusted organisms, and calcareous debris. The people of Toniná intentionally selected and carved these rocks for use as building blocks and bricks on floors, walls, and stairways. At least two coquina flagstones measuring about 1.90 m long were identified in an archeological context most likely associated with carved stelae. Also non-marine carbonate rocks such as a crudely banded travertine and spongy calcareous tufa from recent sediments of freshwater environments surrounding Toniná were used by the Maya as a raw material on walls, columns, reliefs and murals base.

Results of the chemical, mineralogical and micromorphological analysis on plaster and bedding mortars from the walls of Toniná display a slightly interbedded lime of sparry calcite cemented in a highly porous groundmass with silt-to pebble-size of calcareous and siliciclastic rock-crushed aggregates, sand, and soil remains. Lime fabric reveals enclosing quartz grains, granular calcite crystals, and carbonaceous inclusions which may suggest that the lime has been made from a burnt grain-rich limestone with fibrous cement and porous microfabric. WDX analysis in lime lumps of plaster reveal an average amount of 1.37 wt% MgO associated with a limestone source ranging from regular to a magnesium-enriched limestone (1 to 2 wt %). XRF detect a strontium-rich level in the calcite matrix of plaster which is as high as that of fossil shells, tufa, and coquina. Finally, XRD shows that the mean amount of calcite in plaster is 95 wt% and lower amount (2-2.5w%) of siliciclastic minerals: quartz and albite. In contrast, calcite in mortar ranges less than 90.1 wt%. The concentrations of non-carbonate minerals, such as quartz and albite, are higher than those in plaster because mortar incorporates more siliciclastic rock remains, sand and clay.

INTRODUCTION

The archaeological site of Toniná is located to the northeast of the Ocosingo Valley and near the Jataté River (Figure 1), at latitude 16° 54′ 7.49″ N, longitude 92° 00′ 33.15″ W. This ancient city is considered part of the cultural sphere of the southern Maya Lowlands like other larger Maya centers from the Late Classic period such as Piedras Negras, Yaxchilan and Palenque [1]. Toniná is about 64 km south of Palenque; a rival city that Toniná eventually conquered [2].Toniná was a belligerent city-state and one of the most dominant Maya peoples in the Chiapas highlands of Southern Mexico during Late Classic period [3].

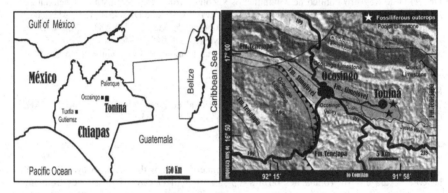

Figure 1. Location of the Toniná archaeological site, Chiapas, México. Left: Geological formations exposed in the Ocosingo Valley.

The site of Toniná was built on a prominent hill that provides scenic quality and a considerable height, as observed in one of the first expeditions to Toniná by Bloom and La Farge in 1926 [4], and subsequent expeditions [2]. Here the Maya took advantage of the surrounding hills in a broken topography of the Ocosingo Valley.

The central architectural core in Toniná is a pyramidal base rising 71 meters on a main square [3]. This massive structure of bricks and stone blocks composed of seven terraces with 13 pyramidal temples gives Toniná the appearance of a military fortress accessed by a long stairway of 260 steps (Figure 2). According to Yadeun-Angulo *et al.* [5], the first four platforms of the structure consist of palaces and temples dedicated to administrative functions, celebrations of seasonal festivities and the worship of the gods of warfare. The fifth platform has the temples honoring the deities of death and the Underworld. The sixth and seventh platforms each have four temples dedicated to the corn festivities and the celestial deities and blood sacrifices, respectively. The unique geometry of Toniná represents a huge labyrinth of temples, palaces and stairways that were built through an episodic construction activity [2].

Figure 2. Two perspectives of Toniná: a front view of the massive structure and top view from the north side (bottom).

The square from which the main pyramidal base emerges represents a plaza with altars, carved monuments, and sculptures. There is also a long court for the Maya ballgame located on a sublevel below the plaza. Bricks and blocks from floors, walls, and staircases in the buildings of Toniná show abundant calcareous fossil material. These materials are strongly associated with several Cenozoic marine fossil-bearing rocks that are exposed in the vicinity of the site. Fossils include abundant bivalves and gastropods, as well as corals and diverse microfossils. Additionally, Bloom and La Farge mentioned that the sculptures from Toniná were carved in yellow, sandy limestones with abundant foraminifera [4].

A previous study of plaster from Toniná based on SEM, PIXE and FTIR have been reported by Mateos-Gonzalez [6]. In the present research, both microscopical and geochemical analyses were applied to characterize the building materials at Toniná, including bricks, blocks, mortar, and plaster. In addition, a paleontological fieldwork in the surrounding Toniná was carried out to identify the fossiliferous rock source.

EXPERIMENTAL-QUANTITATIVE ANALYSIS

All materials analyzed in this research are housed in the Collection of Toniná that belongs to Instituto Nacional de Antropología e Historia (INAH). Plaster and mortar samples were collected from the best-preserved layers on the walls in the temples from the fifth, sixth, and seventh terraces. Superficial layers altered by leaching and organic contamination were avoided. Samples that were analyzed chemically were cleaned to remove the adhering soil particles with dental picks, deionized water, and brushes.

For microscopic analysis, samples were prepared in thin sections. Coquina, sascab, caliche, coral, travertine, tufa, sandstone and limestone samples were collected in and around the archeological site. For comparison, archeological plaster from Palenque was collected from the Temple of the Inscriptions, and restoration plaster from the Group 22 [7]. Limestone samples were collected in the surroundings of the Ocosingo Valley in a zone described as the Ocosingo Limestone from the Tenejapa Formation outcrops [8].

Lime lumps of plaster as described in [9, 10] were measurement with Wavelength Dispersive X-Ray using an Electron Microprobe JEOL JXA8900-R with WDX/EDX combined microanalysis [7]. Plaster samples were mounted in resin with highly polished surface, and analyzed in a high vacuum using a WDX multi-channel with an acceleration voltage of 20 keV, acquisition time of 240 minutes, and elemental current of 2.0×10^{-8} A. Also, SEM microimaging of lime lumps of plaster was performed with JEOL JSM-6360 LV and JEOL JSM-5600 LV equipment, and thin sections of plaster, mortar, and rock were analyzed with stereomicroscope and petrographic microscopy.

X-Ray Diffraction micro-sampling was carried out with Bruker AXS™ D8 Advanced diffractometer with a Bragg-Brentano theta–theta geometry configuration, Cu KR radiation, a Ni 0.5% Cu-K β-filter in the secondary beam, and a one-dimensional position-sensitive silicon strip detector (Bruker, Lynxeye). The diffraction intensity as a function of 2θ angles was measured between 6.5 and 1100, with a 2θ step of 0.039, for 52.8 seconds per point.

X-ray fluorescence analysis in groundmass of plaster, mortar, sascab, and rocks was performed using the portable XRF system SANDRA equipped with Mo tubes and Si-PIN detectors [11]. A current of 0.05 mA and a voltage of 35.0 kV were applied, and two points of analysis were measured per sample, each for a time of 240 seconds.

RESULTS AND DISCUSSION

Rock source

The city core of Toniná lies on a hill with steep sides and a flat top formed by calcareous sandstones, minor siltstones and limestones assigned to a section of the Simojovel Formation exposed in the Ocosingo Valley, dated as Oligocene [8]. The Simojovel Formation is part of the Cenozoic sedimentary cover of the Mountains of the East from the Sierra de Chiapas, as described in [12, 13]. Physiographically, the Mountains of the East correspond to the geological province referred as the Chiapas Reverse-fault, which consist of long, tightly folded mountains separated by very narrow valleys [14]. These mountains and valleys occupy the eastern portion of the Sierra de Chiapas, extending from Ocosingo (in the vicinity of Toniná) to Río Usumacinta, here the mountains decreasing in elevation to the northeast from 2000 to 500 m, as Río Usumacinta is approached.

The bricks and stone blocks in walls, floors and stairways in Toniná comprise sandstones, limestones, boundstones, bafflestones, and coquina flagstones with abundant ostreid shells, gastropods, encrusted organisms, and corals; such rocks and calcareous fossiliferous material are associated with the shallow marine sequences of the Simojovel Formation (Figures 3, 4, 5 and 6). This rocky unit includes lithologies of siltstones, sandstones, bafflestones, framestones, and limestones, which are described as correlated with reef, bioherm, coastal, and shallow marine paleoenvironments [15]. Thus, this is consistent with the nearshore, calcareous fauna found in rocks from Toniná.

The calcareous fossil-bearing beds, however, are not exposed at all sites of Toniná, and there has been no detailed stratigraphic study published for the region of the Ocosingo Valley. Furthermore, the massive architectural structure of Toniná is composed of limestone blocks associated with rocks from the Ocosingo and Quexil Limestones units, as well as the Tenejapa Formation (Paleocene) (see Figure 6.C), which are well exposed surrounding the Ocosingo Valley [8] (see also previous Figure 1).

Coquinas [16] were identified as a stone blocks used on the stairway from the 4th and 5th terrace (Figures 3.A-C). Also two coquina flagstones measuring about 1.90 m long were identified in an archeological context raised on the terraces of Toniná (Figure 3.D-E); these stone blocks were interpreted here as carved stelae.

In the long court of the ballgame were identified bafflestones [16] included as a stone blocks on the floor (Figure 4.A, and B). Microscopical analysis in sample collected on this rock show large fragments of fossil shells, skeletal fragments, calcareous debris, sediment-floored cavities, sparry carbonate cement and rounded siliciclastic particles (Figure 4.C).

Additionally, a boundstone [17] was identified included as stone block on the floor from the 6th terrace (Figure 4.D). A large fragment of calcareous shell surrounded by lithic arenite and white carbonate minerals are observed in thin section of this fossiliferous rock (Figure 4.E). Finally, the Figure 4.F shows a coral specimen recovered on the walls from the 6th terrace (Figure 4.F).

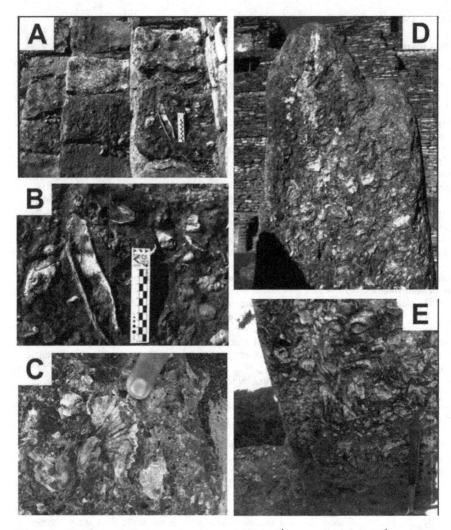

Figure 3. Coquina stone blocks on the stairway from the 4th terrace (A) (B) and 5th terrace(C), respectively. A coquina flagstone raised in the 6th terrace, which probably represents a carved stele (D). Close-up view of previous coquina show abundant fossil shells within the calcareous sandstone matrix (E).

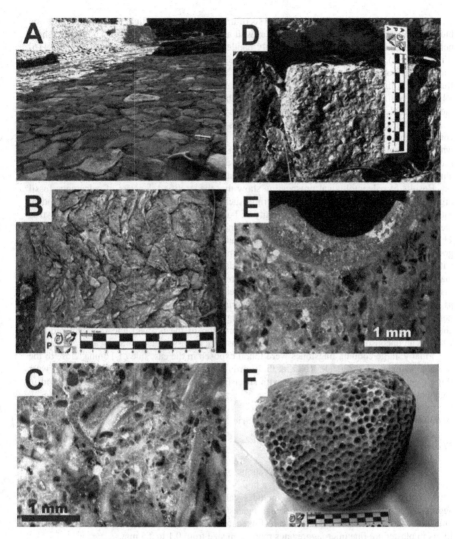

Figure 4. View of the long court of the ballgame (A). Close-up view on the floor of the ballgame show a bafflestone included as a stone block (B). A thin section of previous rock display fossil shells, calcareous debris, sediment-floored cavities, sparry carbonate cement and siliciclastic inclusions (C).Stone block on the floor from the 6th terrace show a boundstone (D). Boundstone thin section show lithic arenite, calcareous shell fragments and white carbonate minerals (E). A coral specimen recovered on the walls from the 6th terrace (F).

On the other hand, microscopical analysis of thin sections of very light, rounded, small stone blocks recovered from the walls, columns, as well as murals and relief bases in Toniná also show non-marine carbonate rocks such as crudely banded travertine and spongy calcareous tufa [16] (Figure 5.A-C), for which the sources are the more recent sediments of freshwater environments surrounding Toniná. These rocks were used by the Maya as a raw material and were incorporated frequently in different architectural and ornamental structures in walls and columns, as well as reliefs and murals supports, such as the Mural of the Four Eras (Figure 5.A). In the present day, the local people refer to such rocks as "chac," a Maya word.

Also, the Figure 5.D shows two stone blocks on the walls from the 7th terrace which are composed by limestone and calcareous sandstone [16]. There are banded siliciclastic minerals, white calcareous inclusions, abundant particles of lithic arenite, and carbonate cement in thin section of sandstone (Figure 5.E). In contrast, the thin section of limestone shows a very white, almost pure carbonate matrix with yellow terrigenous particles (Figure 5.F).

Results of paleontological prospection show that the stratified deposits of calcareous fossil organisms dissolved by an intense weathering, such as mound-like bioherm [16] with abundant ostreid shells, are well exposed in the vicinity of Toniná. Examples of such deposits are shown in the Figure 6.A. This outcrop was identified at the coordinates 16° 54′ 18.6″ north, 92° 00′ 41.59″ west. This unconsolidated, calcareous material often has the texture and color of fine-grained, white dust. The major geochemical composition of this material is calcium carbonate with minor siliciclastic elements. This unconsolidated carbonate dust and sand is an available material in several outcrops in Toniná and the texture and chemical composition is strongly associated with the sascab deposits in the Maya area [18]. The weathering process of fossil shells as occurs in a geological deposit is observed *in situ* in this mound-like outcrop exposed in the north side of Toniná (see previous Figure 6.A). Figure 7 displays a representation of the steps of chemical dissolution over time (following the direction of the arrows) and the final product, which is an unconsolidated calcareous sand and dust that are associated with sascab.

Finally, large coquina flagstones were found exposed in the east side of Toniná, close to the seasonal tributary of the Jataté River, at the coordinates 16° 54′ 7.26″ north, 92° 00′ 5.57″ west (Figure 6.B). Such coquina flagstones with sheet-like forms are associated with those found in archeological context in Toniná.

Bedding Mortar, Plaster and Lime Microfabric

Plaster:

The microanalysis of oriented thin sections of plaster from Toniná shows a recognizable layer of slightly interbedded white lime in a highly porous groundmass and other coarser layer of silt to pebble-size filler mineral that includes calcareous and large siliciclastic crushed-rock aggregates (Figure 8.A). Calcareous aggregates are associated with incompletely burned limestone and siliciclastic aggregates are linked to sand, clay, and soil remains. The mineral fillers in plaster are fine microaggregates ranging in size from 0.1 to 2.5 mm.

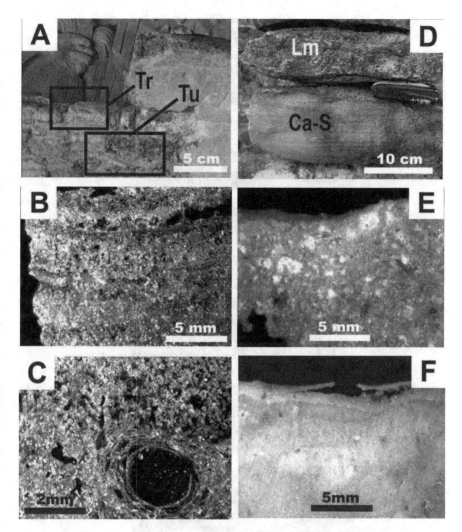

Figure 5. Two bricks on the wall from the Mural of the Four Eras are composed of travertine (Tr) and tufa (Tu) (A). A crudely banded travertine (B), and spongy calcareous tufa (C) as seen both in thin sections, note the high porosity microtexture in tufa associated with a distinctive lightness of this rock known as "chac." Figure D shows two stone blocks on the walls from the 7th terrace which are composed by limestone (Lm), and calcareous sandstone (Ca-S). Figure E and F show a sandstone and limestone thin sections, respectively.

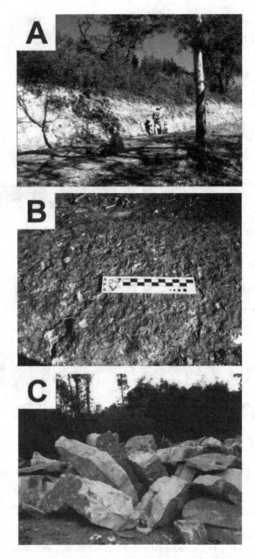

Figure 6. View of calcareous, fossiliferous section exposed on the north side of Toniná (A). This outcrop is composed of dissolved shells, and unconsolidated carbonate sediments, sandstone and clay. A large coquina flagstone exposed in the vicinity of Toniná (B). Distinctive limestone from the Tenejapa strata exposed near to the Ocosingo Valley (C).

Figure 7. Representation of the dissolution pattern over time (following the arrows) of fossil shells as occurs in the section exposed on the north side of Toniná. The final product is an unconsolidated, calcareous sand and dust that are associated with sascab.

The carbonate matrix shows vesicles and lime lumps (Figure 8.B); as well as shrinkage microfractures, flexible grain deformations, coarse neospar, and desiccation structures associated with decay of the plaster matrix that increases porosity and calcium dissolution (Figures 8.D- F). Leaching, coarse neospar, and large process of recrystallization are observed in plaster from the North Palace (Figure 8.F). Coal lumps are also observed, this is evidence of the slake lime production by burning rocks (Figure 8.C). Plaster from the Palacio de Las Grecas shows a surficial layer of paint with brownish color highly altered by pigments oxidation and intense recristallization by decay process (Figure 8.D). On the other hand, plaster from the Mural of the Four Eras shows large, red particles of Fe oxide incorporated in more internal layers of the lime mixture below the thin layer of red paint (Figure 9.A). The large particles of Fe oxides are remnants of hematite pigments. Here, the white lime groundmass interbedded with mineral fillers is clearly observed. In contrast, mortar from the same place shows a coarser microfabric than plaster with lower concentrations of white lime that is dispersed throughout the cement matrix of the mortar (Figure 9.B).

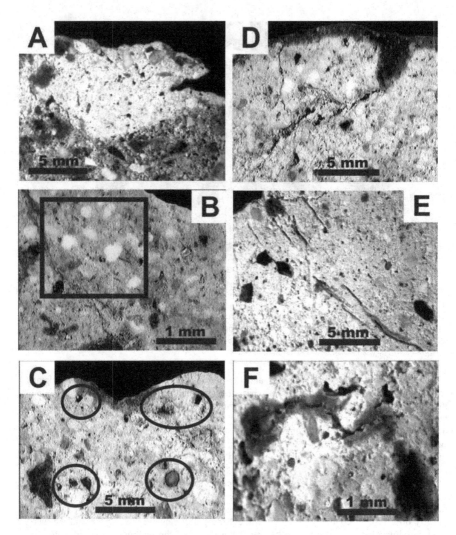

Figure 8. Plaster from the Mural of the Four Eras (A). Lime lumps of plaster from the previous structure indicate by the rectangle (B). Plaster from the Palacio de las Grecas (C), circles show coal lumps. Plaster of the North Palace from the 5th terrace (D). Plaster recovered from a temple indet. in the 6th terrace (E). Plaster recovered from a palace indet. in the 4th terrace (F).

Mortar:

In general terms, mortars of Toniná shows a highly cemented matrix of calcareous clay, the crushed rock aggregates in mortar, ranging in size from 0.5 to 12 mm, are larger and found in higher concentrations than those found in plaster (Figures 9.B-F). The mineral fillers in mortar show a different sedimentary source, for instance, white calcareous inclusions are observed, also yellowish to brownish dust from siliciclastic sources with recognizable inclusion of feldspars and detritical quartz, as well as other reddish particles that indicate clay content; the Figure 9.E summarizes the above described.

Mortar from the Mural of the Four Eras show abundant siliciclastic aggregates (Figures 9.B), and mortar from a temple indet. in the 6th terrace shows a yellow to brownish lithic arenite, calcareous grains, and large limestone aggregates (Figures 9.E, and F). Here is also observed a finely portion of rounded particles interbedded with larger mineral fillers, which is related with the texture of sand (Figure 9.E).

Lime microfabric:

Additionally, a microstructural examination of plaster samples from the Mural of the Four Eras and Palacio de Las Grecas were made. So the results in lime plaster from the Mural of the Four Eras (Figure 10.A) show recrystallized calcite crusts and granular calcite crystals with high interparticle porosity that may suggest that the lime was made from a burnt grain-rich limestone with micrite/sparry cement and porous microfabric [17]. Plaster from the Palacio de Las Grecas also shows two different porous microtexture: a coarse-grained calcite crystals poorly packed linked to sparry calcite (Figure 10.B), as well as elongated carbonate grains linked to a tufa within the crystalline frame of calcareous clay source (Figure 10.C).

WDX Measurements of Lime Lumps

The results of WDX analysis are summarized in Table 1. Lime lumps of plaster from Toniná show low concentrations of MgO ranging from 0.65 to 2.80 weight percentage (wt %) while the CaO content ranges from 34.2 to 36.9 wt %. The average MgO content is 1.38 wt % and the average CaO content is 35.6 wt %. Limestone samples recovered from the Tenejapa strata exposed near Toniná also show poor MgO content (between 1.05 and 1.39 wt %). Note also that the MgO contents for standard limestone in sample S20 are significantly lower than those from the standard dolomite in sample S50. These results for MgO concentrations in plaster from Toniná may suggest a limestone ranging from regular to a magnesium-enriched limestone as described by the American Geological Institute [19]. There are significant variations in the stoichiometric composition of magnesium and calcium weight percentages in carbonate rocks (Boggs, 2009). According to Neuendorf *et al.* [19] a typical limestone has MgO content less than 1.1 wt% while magnesium-enriched limestone shows MgO content ranging from 1.1 to 2.1 wt %.

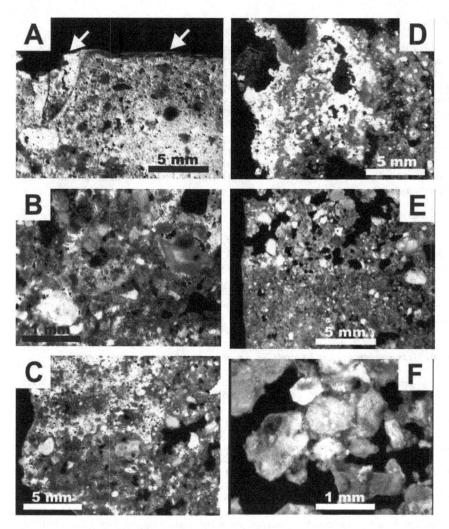

Figure 9. Plaster collected from the Mural of the Four Eras (A), the thin layer of red paint on the top is shown by arrows. Mortar from the same place shows siliciclastic mineral fillers in the lower layer (B). Mortar collected from the Palacio de las Grecas (C). Mortar recovered from the North Palace (D). Mortar recovered from a temple indet. in the 6th terrace (E). Close-up view to previous sample shows the calcareous rock-crushed aggregates (F).

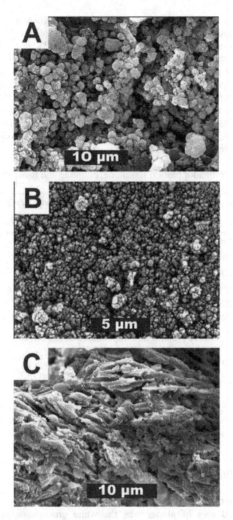

Figure 10. Electron micrograph showing a surficial layer of granular calcite crystals and porous microtexture in lime plaster from the Mural of the Four Eras (A). Microfabric of coarse-grained calcite crystals in lime plaster collected from the Palacio de Las Grecas (B). Elongated calcite crystals in a spongy groundmass of plaster from the previous structure (C), this microtexture are associated with tufa.

In contrast, plaster samples from Palenque show a higher content of MgO than Toniná plaster: in samples P10 and P11 the detected concentrations of MgO are 11.16 and 12.35 wt %, respectively, and concentrations of CaO are 21.05 and 22.29 wt %, respectively. Additionally, plaster from restoration material from Palenque (sample PR1) also shows poor MgO content with 0.29 wt % and CaO content of 39.02 wt %.. Additional MgO/CaO values from Palenque plaster that contrast with the plaster of Toniná have been recently reported in [7].

In addition, significant concentrations of trace elements are also detected in lime lumps: elements ranging from Na to Sr in the form of oxides compounds, such as P_2O_5 (1085 µg/g), FeO (1013 µg/g), and SrO (1687 µg/g), among others (see Table 1). Concentrations of SrO greater than 1000 µg/g detected in Toniná plaster are not observed in rocks; however SrO concentrations observed in fossil shells are on average 4120 µg/g.

Table 1. WDX results for plaster, limestone, dolomite and fossil shell samples

Sample	Type	Na$_2$O µg/g	Al$_2$O µg/g	SiO$_2$ µg/g	P$_2$O$_5$ µg/g	K$_2$O µg/g	MnO µg/g	FeO µg/g	ZnO µg/g	SrO µg/g	MgO wt%	CaO wt%
T6	plaster	780	830	-	1010	200	390	910	230	1850	0.89	35.8
T9	plaster	210	2560	-	840	300	580	1520	100	1480	0.65	36.5
T12	plaster	590	590	-	760	940	160	850	-	1130	1.01	34.9
T11	plaster	690	310	-	1480	140	80	1020	-	2280	2.80	34.2
T25	plaster	550	390	-	1120	600	120	1050	-	1630	1.09	35.2
T26	plaster	120	40	-	1300	370	100	730	21	1750	1.85	36.9
P10	plaster	430	410	680	770	90	60	420	-	3740	12.4	21.9
P11	plaster	-	680	-	960	20	110	710	-	3670	11.2	22.3
PR1	restoration	90	280	-	610	-	-	-	210	60	0.29	39.0
T47	limestone	-	20	-	480	30	-	740	-	120	1.05	38.7
T48	limestone	-	340	-	950	90	-	441	-	-	1.39	39.0
S20	std limestone	-	-	-	770	350	-	300	-	-	0.01	40.1
S50	std dolomite	1690	30	-	530	-	190	220	260	-	23.1	30.4
F2	Fossil shell	580	120	-	1010	20	20	835	710	4120	0.23	39.0

XRF measurements of Plaster, Rocks, and Sascab

Plaster samples from the walls of fifth, sixth, and seventh terraces in Toniná were analyzed with XRF. The analyzed area of plaster was the white groundmass that appears slightly interbedded with a coarser layer of mineral fillers. XRF measurements give a basic guideline to characterize the composition of plaster, identifying both major elements and impurities. These impurities also allow a comparison of plaster with carbonate minerals from rocks. XRF measurements in the plaster detected high intensities of Ca, Mn, Fe, and Sr (Figure 11). Ca levels are nearly constant in plaster; nevertheless, changes in the levels of Mn, Fe, Zn, and Sr indicate variations in the impurities present in the carbonate matrix of plaster. In coarser plaster there are increases in the Mn and Fe levels associated with the major concentration of mineral fillers. Also the lower amounts of dissolved carbonate shows this variation in Mn, Fe and Sr in sascab, fossil shell, limestone, calcareous tufa, and coquina. So there is less Sr content in limestone than in the plaster groundmass. In contrast, the Sr levels in fossil shells and coquina are as high as that of plaster. This is consistent with the presence of carbonate minerals with a biogenic source that show strontium high-content as relict of aragonite, which can be the main composite of calcareous shells [17]; however, aragonite is transformed in calcite by diagenetic process in most geological deposits [16]. Thus, aragonite is no detected by XRD in mortar or plaster of Toniná, even so, significant concentrations of Sr have been found. Strontium could be a useful marker of the presence of dust of calcareous shell in plaster and mortar; however, we suggest that it is preferable to wait applying additional measurements to confirm the chemical fingerprints of dust of calcareous shell in plaster and mortar.

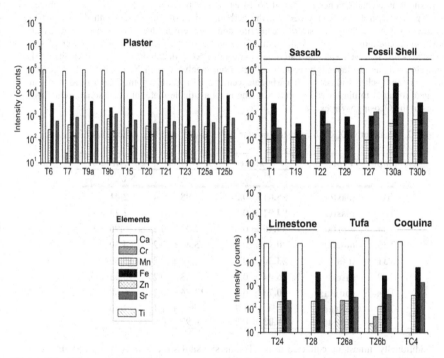

Figure 11. Diagnostic XRF measurements obtained for plaster, sascab, coquina, fossil shells, limestone, and tufa.

Mineral characterization by quantitative XRD

Table 2 shows the results of XRD analyses of plaster and mortar from Toniná, as well as rocks and calcareous sediments collected from the outcrops in the vicinity of the archaeological site. The concentration of calcite in plaster is greater than 94.91 and less than 95.61 wt %, with minor amounts of siliciclastic minerals such as quartz, found in sand [17], in concentrations between 1.98 and 2.41 wt %, as well as albite, associated with clay [16], in concentrations between 2.41 and 2.45 wt %. The amount of calcite in mortar is greater than 89.1 and less than 90.10 wt %. The concentrations of non-carbonate minerals, such as quartz and albite, in mortar are higher than those in plaster because mortar contains many siliciclastic rock remains, sand and clay. Quartz and albite are clearly associated with the texture of mortar that incorporated siliciclastic mineral fillers in greater quantities than plaster. The white sand and dust collected from mound-like deposits in Toniná and associated with sascab [18] are also carbonate composite with a calcite concentration of 86.44 wt %. As shown in the XRD values, sascab can have high concentrations of clay because the carbonate deposits are mixed with the soil when they are dissolved by the processes of weathering and pedogenesis that form sascab; such sascab from Toniná contains 10.33 wt % albite and 3.23 wt % quartz. Caliche, a surface layer of soil encrusted with dissolved calcium carbonate exposed in very dry and weathered deposits [16], found close to sascab deposits was also collected and analyzed with XRD (Table 2). Because caliche is sandy clay with calcium carbonate cement [17], probably the Maya used it as an aggregate material in plaster and mortar. Caliche is an available material in seasonal creeks and ravines in the Ocosingo Valley.

Table 2. Quantitative XRD for plaster, mortar, tufa, sascab, coquina, caliche and limestone.

Sample	Type	Calcite	Quartz	Albite
		CaCO$_3$	SiO$_2$	NaAlSi$_3$O$_8$
T6	plaster	95.61 ± 0.2	1.98 ± 0.4	2.41 ± 0.3
T25	plaster	94.91 ± 0.6	2.84 ± 0.2	2.25 ± 0.3
T34	mortar	89.10 ± 0.6	2.00 ± 0.2	8.90 ± 0.3
T35	mortar	90.10 ± 0.6	5.40 ± 0.2	4.50 ± 0.3
T31	tufa	83.30 ± 0.7	1.20 ± 0.2	15.50 ± 0.5
T29	sascab	86.44 ± 0.7	3.23 ± 0.2	10.33 ± 0.7
TC4	coquina	80.73 ± 0.5	12.46 ± 0.3	6.80 ± 0.4
T42	caliche	4.56 ± 1.7	59.4 ± 1.1	36.0 ± 1.1
T45	limestone	99.03 ± 0.2	0.97 ± 0.2	-

Additionally, limestone collected in the Toniná site shows a calcite content of 99.03 wt % with low quartz content. In contrast, calcareous tufa and coquina collected in the archaeological site show a high amount of non-carbonate minerals; tufa contains 15.50 wt % albite, while coquina contains 12.46 wt% SiO$_2$ (quartz). The amount of albite present in calcareous tufa is also associated with the pedogenesis process [17], tufa is formed by the precipitation of carbonate minerals in the presence of organic material and soil sediments at warm temperature in water bodies [16].The significant amount of SiO$_2$ in coquina is linked to the siliciclastic material (mostly sand) incorporated in the calcareous shells and carbonate debris that form the biogenic rock that the Maya used in the construction of Toniná.

CONCLUSIONS

The research of ancient building materials rarely takes into account the presence of fossils in lithic materials. However, the present study benefits from insights provided by the paleontological fieldwork carried out at the site of Toniná. As shown in [7, 20], such paleontological prospection clarifies the main features of bricks, stone blocks, and flagstones enriched with fossil material found in the archaeological context. The fieldwork is also helping to identify the rock sources in geological deposits exposed in the Ocosingo Valley. The data present in this work represent the first approach that integrates the description of fossiliferous rocks and sediments used by the Maya people in Toniná. Therefore, this paper documents the variability of the lithic raw material used by the builders of Toniná. The Maya intentionally selected this material, likely considering the basic properties of rocks and sediments, such as hardness, size, thickness, porosity, and texture, as well as resource availability. This confirms that the people of Toniná adapted the local rock resources for their own uses, exploring, selecting and using the lithic raw material with expertise.

ACKNOWLEDGEMENTS

Authors thank to Yolanda Hornelas from ICMYL-UNAM for SEM imaging. We also are deeply indebted to Julisa Camacho and the archeologist Margarita Reyes, as well as the people of INAH from the archeological site of Toniná, for field assistance. This project was funded by grant CONACYT-346471/239885 and Ciencias Biológicas-UNAM postgraduate program. Field work was supported in part by grant UNAM-PAPIIT IN106011 while the experimental analysis was performed with support from UNAM-PAPIIT IN403210 and CONACYT 131944 MOVIL II endowment.

REFERENCES

1. L. Pescador-Cantón, Toniná, la Montaña Sagrada de los señores de las serpientes y los jaguares, in: *Las culturas de Chiapas en el periodo prehispánico,* D. Segota (ed.), Conaculta, Mexico, 2000, 245-281.
2. P. Becquelin, C. Baudez, Toniná, une cité maya du Chiapas (Mexique), vol 6, tomos I, II y III, *Collection Études Mesoamericaines et Centroamericaines,* México, 1982.
3. C. Baudez, E. Taladoire Toniná, une cité maya du Chiapas (Mexique), vol 6, tomo IV, *Collection Études Mesoamericaines et Centroamericaines,* Mexico, 1990.
4. F. Bloom, O. La Farge, *Tribes and Temples, a record of the expedition to middle America conducted by the Tulane University of Louisiana in 1925,* The Tulane University of Louisiana, New Orleans, 1926, 536.
5. J. Yadeun-Angulo, J.I. González-Manterola, P. Oseguera-Iturbide. *Toniná: el Laberinto del Inframundo.* Gobierno del Estado de Chiapas, México, 1992, 127.
6. F. Mateos-González, *Toniná: la pintura mural y los relieves en estuco. Técnica de manufactura.* Instituto Nacional de Antropologia e Historia, INAH y Centro de

Investigaciones Humanísticas de Mesoamérica y el estado de Chiapas, UNAM, Colección científica serie Arqueología, México, (1997).104.

7. F. Riquelme, J. Alvarado-Ortega, M. Cuevas-Garcia, J.L. Ruvalcaba-Sil., C. Linares-López, *Journal of Archaeological Science* 39 (2012) 624-639.
8. H. Martinez-Amador, E. Cardoso-Vásquez, E. Ramirez-Tello, H. Membrillo-Ortega, V. Minjares-Rivera, *Carta geológico-minera Las Margaritas, Chiapas. (E15-12 D15-3),* Thematic map and Geological descriptions, SGM-INEGI, México, 2006.
9. A. Bakolas, G.A. Biscontin, A. Moropoulou, E. Zendri, *Thermochimica Acta* 269-270 (1995) 809-816.
10. L. Barba, J. Blanca, L. Manzanilla, A. Ortiz, D. Barca, G.M. Crisci, D. Miriello, A. Pecci, *Archaeometry* 51 (2009) 525–545.
11. J.L. Ruvalcaba, D. Ramírez, V. Aguilar, F. Picazo, *X-ray Spectrometry* 39 (2010) 338-345.
12. F.K.G. Mullerried,. *Boletín de la Sociedad Geológica Mexicana,* Tomo XIV (1949). 73-100.
13. K.M. Helbig, *La cuenca superior del río Grijalva,* Instituto de Ciencias y Artes de Chiapas, México, 1964, 247.
14. J.J. Meneses-Rocha, Tectonic evolution of the Ixtapa graben, an example of a strike-slip basin in southeastern Mexico: implications for regional petroleum systems, in: *The Western Gulf of Mexico Basin: Tectonics, Sedimentary Basins, and Petroleum Systems,* edited by C. Bartolini, R.T. Buffler, A. Cantú-Chapa, American Association of Petroleum Geologists Memoir 75, Tulsa, Oklahoma, 2001, 183-216.
15. R.C. Allison, *The Cenozoic Stratigraphy of Chiapas, Mexico, with Discussions of the Classification of the Turritellidae and Selected Mexican representatives.* Unpublished Ph.D. thesis, University of California, Berkeley, 1967.
16. S. Boggs. *Petrology of Sedimentary Rocks.* Cambridge University Press, Cambridge UK, 2009, 600.
17. E. Flügel, *Microfacies of carbonate rocks. Analysis, interpretation and application,* Springer-Verlag, Berlín, 2004, 984.
18. E.R. Littmann, *American Antiquity* 24 (1958) 172-176.
19. K.K.E Neuendorf, J.P. Mehl, J.A. Jackson, *Glossary of Geology,* American Geological Institute, Alexandria, Virginia, 2005, 779.
20. M. Cuevas-García, J. Alvarado-Ortega, *Estudio Arqueológico y Paleontológico de los Fósiles marinos que proceden del sitio de Palenque, Chiapas.* 1st Field Work Season Report 2008, Instituto Nacional de Antropologia e Hisotoria INAH, Mexico, 2009, 54.

Conservation Studies

Mater. Res. Soc. Symp. Proc. Vol. 1374 © 2012 Materials Research Society
DOI: 10.1557/opl.2012.1386

How to Make a Latex Rubber Sculpture?
Manufacturing Studies that Improve the Creative Work of an Artist

Nora A. Pérez[1], José L. Ruvalcaba[2], Claudio Hernández[3], César Martínez[4]

[1]Instituto de Investigaciones en Materiales, Universidad Nacional Autónoma de México, Circuito Exterior s/n, Ciudad Universitaria, México, D.F. 04510, Mexico.
e-mail: norari.perez@gmail.com

[2]Instituto de Física, Universidad Nacional Autónoma de México, Circuito Investigación Científica s/n, Ciudad Universitaria, México, D.F. 04510, Mexico. e-mail: sil@fisica.unam.mx

[3]Laboratorio de Restauración, Museo Universitario Arte Contemporáneo, MUAC-UNAM, Zona Cultural Universitaria, Insurgentes Sur 3000, México, D.F. 04510, Mexico.
e-mail: claudiohhddzz@yahoo.com

[4]Departamento de Evaluación del Diseño en el Tiempo, Universidad Autónoma Metropolitana, Av. San Pablo No.180, Col. Reynosa Tamaulipas, México, D.F. 02200, Mexico.
e-mail: cmartinez62@yahoo.com

ABSTRACT

Contemporary Art has the characteristic of being made with a wide diversity of materials. In this plastic age many of the artists employ polymers to create their works, but do not consider the degradation that their art will suffer eventually.

This work presents the studies performed for improving the manufacture of a series of sculptures made of latex rubber that belong to the Museo Universitario Arte Contemporáneo (MUAC), UNAM, in Mexico City. These sculptures made by César Martínez are blow up and deflated continuously during their exhibition. Techniques such as Fourier Transformed Infrared (FT-IR), X-ray fluorescence (XRF) and Raman spectroscopies were used for characterizing the manufacturing techniques of the artist. Dynamical Mechanical Analysis (DMA) was carried out to correlate the mechanical properties with the raw materials.

These analyses provide a comprehensive understanding of the material and the main factors that affect the degradation of the pieces. The combination of these studies made possible to suggest a new methodology to the artist in order to improve the quality and therefore enlarge the lifetime of his work.

INTRODUCTION

At the end of the nineties, the Museo Universitario Arte Contemporáneo (MUAC) of the Universidad Nacional Autonoma de Mexico (UNAM) requested to the artist César Martínez Silva [1] the production of an artwork for the collection of the museum. The artist decides to produce a piece made of latex rubber called *El desgaste de la clase media mexicana* that was composed by three sculptures; a standing man, a laying woman on the floor and a baby (see figure 1). These sculptures made entirely by the artist evoke thoughts about the breathing process by making the sculptures blow up and deflate when the visitor stands in front of the pieces, the mechanism employs an air gun, a flexible tube and a movement detector.

Figure 1. César Martínez Silva, *El desgaste de la clase media Mexicana*, baby latex rubber sculpture inflated (left), deflated (right).Collection of MUAC UNAM.

When these pieces were intended for exhibition ten years later the latex was seriously degraded, it was a rigid material with fractures and since the essential functionality of the piece was lost (the mechanism of blowing and deflating) reproductions were created following the evolution of the artist's manufacturing process. The artist continued producing a series of sculptures of the same type that he called *Sucesos Escultóricos*.

The Artist's manufacturing process

The artist himself developed the method of producing the sculptures which begins with making the fiber glass casts modeled from live human beings; afterwards the artist mixed natural rubber latex with a vulcanizer solution, both of an unknown composition. He used silicon dioxide as filler; employing different brands depending on the year he made the sculptures finally he painted the latex with an outdoors black paint. He applied 20 layers of latex on the glass fiber casts; each layer is heated between 40°C and 60°C during 20 minutes. Since 2002 he placed a textile between the 12 and 15 layers as structural support. The textiles were different for each piece and included tights and cheesecloth.

The aim of this study was to characterize the material, determine the alteration process and through the results of these studies improve the formula made by the artist. Finally, the museum requested to propose conservation strategies for the piece's exhibition and storing. In this study, the artist's participation was very important since he wanted also to improve the properties of their sculptures and the manufacturing procedure.

METHODOLOGY

Three samples (see figure 2) were studied that had different temporality (1998, 2002 and 2010), and therefore had different manufacturing processes since the artist had changed his methodology through the years. There is a differentiation between the smooth side of the samples (side in direct contact with the cast) and rough side (top part of the latex layers). Also the raw materials were studied in order to establish the alteration degree. After evaluating the artist's samples, new latex samples were made with the raw materials provided by the artist to compare with the latter ones; the only difference is the use of carbon black instead of black paint to color the latex.

The methodology of study consisted first in a general microscopic observation in order to assess the material differences between the samples; the samples were observed through a stereoscope with a white light lamp. Afterwards, in order to evaluate the morphology of the surface a JEOL JSM-5600 Low Vacuum Scanning Electron Microscope was employed. For the analysis of the inorganic components of the sculptures such as the filler and the vulcanizer the SANDRA X-ray fluorescence equipment developed in the Physics Institute of the National University (UNAM) was used. A Mo X-ray tube with 25kV and 0.150mA was used, each measurement lasted 180s and two different regions for each side of the sample were studied.

Figure 2. Sculpture samples from different years. The images marked with an asterisk show the rough side. Year 1998 (a, a*), year 2002 (b, b*), year 2010 (c, c*).

Afterwards, Infrared (FTIR) and Raman spectroscopies were applied for the study of the organic matrix. Spectra were recorded on a Nicolet 6700 Thermo Electron Corp spectrophotometer within the range 400 a 4000cm^{-1} at room temperature with a resolution of 4cm^{-1} in the case of the FTIR study. A Delta Nu Inspector spectrometer with a 785nm laser of 120mW was employed for the Raman study within the range 200 a 2000cm^{-1} with a resolution of 8cm^{-1}. Finally, a Dynamical Mechanical Analysis was carried out to evaluate the decrease of mechanical resistance due to the elongation process for determining the degree of functionality of the pieces. The TA Instruments DMAQ800 in tension mode was used, with a ramp of 0.1N/min up until the breaking point of the plastic at room temperature (25°C). The test was performed three times for each sample.

RESULTS AND DISCUSSION

Microscopic examination

The microscopic observation provided information about the homogeneity of the material, the light reflection on the sample helped as an indicator of the quantity of the fumed silica particles. In figure 3 it is evident the differences in the distribution of the filler, in the first picture that corresponds to the 1998 sample we can observe that the particle size of the silica is fine and it is well distributed through the material, while in the others, the amount of silica is higher being the highest in the 2010 sample; also the distribution of the filler is poor since the quantity has increased. This difference on the quantity of filler affects the visual appearance of the sculpture and instead of reinforcing the material it decreases the resistance, since it extends the volume of the system and the porosity causing early fractures. This increase in the amount of filler is because the artist changed to fumed silica of smaller particle size but maintained the proportion of the first formula.

A low quantity of pores was observed in the 1998 sample, while in the 2002 sample there is an increase in the pores but also in fractures especially in the textile's fibers direction indicating that this type of support isn't the best for the piece, since it will fail easier with the mechanical efforts it is subjected to. In the 2010 sample the increase in the number of pores is directly related with the type of textile employed which has a wider separation between fibers which keeps the latex from forming a uniform layer.

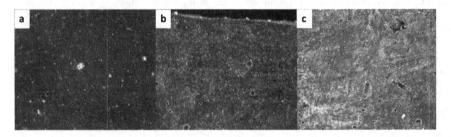

Figure 3. Microscopic examination of the rough side of the samples. 1998(a), 2002(b), 2010(c).

From the SEM images (see figure 4) of the latex surface it is observed the distribution of the filler. The same behavior that was observed in the stereoscope is reproduced here, for the 2010 latex the fumed silica particles are no longer on the polymer but are actually totally deposited on the polymer surface and clusters are formed. About the degradation of the polymer, in the 1998 latex the fractures are evident while there in the 2002 there is no evidence of fractures on the polymer; in the 2010 sample due to the filler it is not possible to observe the polymer. On the images of the cross sections it is more evident the degree of degradation, on the first sample the polymer is totally fractured and has lost its plastic appearance; in the 2002 sample it is observed the textile yarn at the top left of the image, the sample is fractured but it still has a plastic appearance which allows the sculpture to have a good mechanical performance still. In the image that corresponds to the 2010 latex it is observable the layer of fumed silica at the top of the sample, the polymer does not exhibit fractures but it has pores that are distributed along the textile which eventually can accelerate the degradation of the polymer.

To establish the level of degradation new samples were made to compare with the latter ones. In figure 5 it is evident the good appearance of the cross section of the new materials in comparison to the artist's samples. The amount of filler used for these new samples was of 1%, in the sample that contains carbon black a change in the texture is noticeable specially because the filler is well mixed with the polymer and it's not on the surface as in the case of the SiO_2; the white stain that is observed is due to a fumed silica cluster that was broken at the moment of cutting the sample.

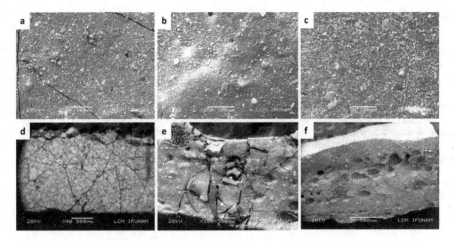

Figure 4. SEM images of the latex samples. Surface images(top): 1998(a), 2002(b), 2010(c). Cross sections (bottom): 1998 (d), 2002(e), 2010(f).

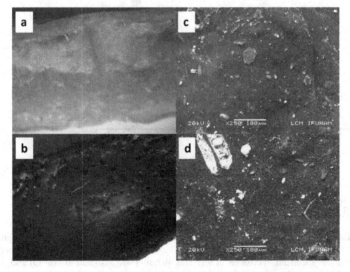

Figure 5. New latex samples. Microscopic images of the surface (left) and SEM images of the cross sections (right). Latex and SiO_2 as filler (a) and (c), Latex with carbon black and SiO_2 (b) and (d).

X-ray fluorescence analysis

The analysis performed on the raw materials (see figure 6) show that the latex solution is not only the isoprene solution but it also contains sulfur and zinc; this last element must be an activator for accelerating the vulcanizer process [2]. The vulcanizer solution has a similar quantity of sulfur and a smaller amount of zinc in comparison to the latex solution, the solution presents potassium and calcium. The fumed silica analyzed has a large amount of calcium impurities, while the main inorganic component in the black paint is also calcium.

In figure 7 it is observable the change on the intensities for the latex samples. The sulfur signal decreases as the polymer is older as expected, since with ageing the sulfur links breaks producing sulfur gases. The signal is weaker in the smooth side since this side is in direct contact with air and light. The silicon content remains constant trough time for all samples; a small increase in the signal of the 2010 sample is in accordance with the microscopic results. The amount of calcium and potassium decreases dramatically while the zinc signal increases as the polymer ages indicating that the ion is less inhibited as a consequence of the polymer matrix degradation.

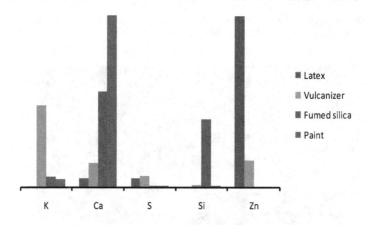

Figure 6. Normalized X-ray intensities (a.u.) of the raw materials.

Infrared and Raman spectroscopies

The infrared (FTIR) and Raman spectra obtained from the latex solution (see figure 8) indicate that the polymer is a natural rubber [3]. The C=C stretch at 1665cm[-1] is a prominent band in the Raman spectrum, the bands at 997 and 492cm-1 are assigned to v(S=O) and v(S-S) [4], respectively, since it has already been identified the presence of sulfur in the latex solution by XRF. The artist's samples have been studied only by IR spectroscopy due to the limiting factor of the black paint he used which causes strong fluorescence in the Raman spectra.

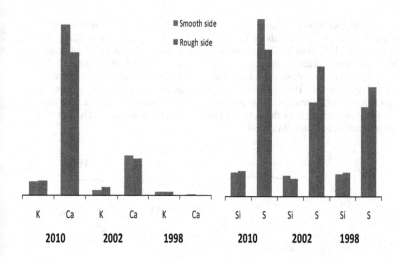

Figure 7. Normalized X-ray intensities (a.u.) of the artist's samples.

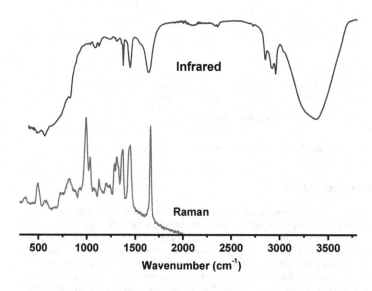

Figure 8. FTIR and Raman spectra of the raw latex solution.

All the latex samples (see figure 9) have bands at 2962, 2923, 2853 cm^{-1} corresponding to $\nu_{as}(CH_3)$, $\nu_s(CH_2)$ and $\nu_s(CH_3)$. The bands at 1436, 1372 cm $^{-1}$ are assigned to $\delta_{as}(CH_3)$ and $\delta_s(CH_2)$ [4]. The bands at, 1731 and 3330 cm^{-1} are assigned to carbonyl and hydroxyl groups respectively, this bands are present in the older samples due to the oxidation process that degrades the polymer [5]. It is observable the intensity decrease of the band at 1021cm^{-1} which is assigned to the $\delta(C\text{-}H)$ of the -S-CH=CH$_2$ functional group; as the polymer ages the band at 1092cm^{-1} assigned to a sulfoxide group [4] is observable in the 2002 and 1998 spectra. The band at 470 cm^{-1} $\delta_{as}(Si\text{-}O\text{-}C)$ decreases [4] as the polymer gets older as the polymer matrix degrades and the molecular interaction with the filler breaks.

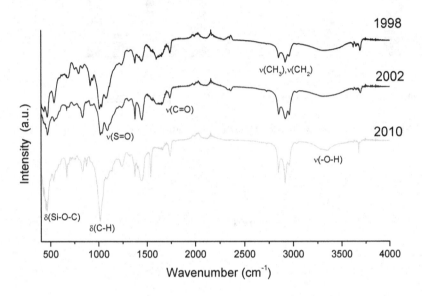

Figure 9. FTIR spectra of the artist's latex samples.

Mechanical dynamical analysis (DMA)

All three polymer samples exhibited the expected behavior of an elastomer. A linear regression was used on the elastic zone of the curve to calculate the value of modulus of elasticity. The average modulus for the 1998 samples was 0.457 MPa/mm, for 2002 of 0.564 MPa/mm and for 2010 was 0.587 MPa/mm. The three values are similar which indicates that the elastic modulus of the material is 0.536 MPa/mm, but as the material gets older it is not able to withstand the same stress obtaining that the maximum force for the newest sample with 1.78MPa and the minimum is 0.50MPa for the 1998 sample. The strain values were 15% for the 1998 latex, 200% for the 2002 sample and 100% for the 2010 sample; this variation in the strain value for the 2010 sample shows the negative effect of the filler excess. From this, it is still recommended to employ an eight year old material, but it wouldn't be advisable to use an older material made twelve years ago since it will not endure long the continuous stretch the sculpture is subjected to.

New samples were made with different types of fillers, the results obtained for the DMA (see figure 10) shows that the best formula is the one that includes carbon black and fumed silica as fillers, showing more than 300% of strain, the latex that has tights as structural support has a 270% of strain. The maximum stress is much lower than the artist's samples but the strain values are higher, due to the sculptures mechanical performance it is preferable to have a greater strain value.

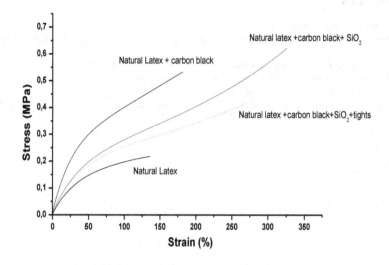

Figure 10. DMA results of the new latex samples.

CONCLUSIONS

The main degradation factor of the sculptures is an oxidative degradation process that affects the polymer; but the mechanical characteristics are affected by the latex homogeneity which is directly related to the amount and type of filler, and also to the type of textile employed. The recommendations given to the artist were to add a phenolic antidegradent to the formula in order to prevent oxidation. A second recommendation was to carefully measure the filler and always take into account the particle size of the raw material, and finally according to the experiments we strongly recommended the use of carbon black as filler since it will also color the latex avoiding the use of other materials. In respect to the museum request a proposed conservation strategy was to store the sculptures partially filled with an inert gas for preserving its mechanical properties.

ACKNOWLEDGMENTS

Authors acknowledge the Museo Universitario Arte Contemporáneo (MUAC) of UNAM, for the collection studies and the supports of UNAM PAPIIT IN403210 project as well as the CONACYT Mexico grant 131944 to carry out this study. This research has been performed in the frame of the Non Destructive Study of the Mexican Cultural Heritage ANDREAH network (www.fisica.unam.mx/andreah).

REFERENCES

1. http://www.martinezsilva.com/
2. K. Nagdi, *Rubber As Engineering Material:Guideline For Users*, Hanser Publ., Munich,1993, 12.
3. P. J. Hendra, K. D.O. Jackson, *Spectrochimica Acta* 50 A 11 (1994) 1987.
4. G. Socrates, *Infrared and Raman Characteristic Group Frequencies, Tables and Charts*, John Wiley & Sons, Chichester, UK, 2001.
5. J. F. Rabek, *Polymer Photodegradation, Mechanisms and Experimental Method*, Chapman & Hall, London, 1995.

Mater. Res. Soc. Symp. Proc. Vol. 1374 © 2012 Materials Research Society
DOI: 10.1557/opl.2012.1387

Decay Degree determination of Archaeological Shell Objects from the Great Temple of Tenochtitlan, using a Visible Light Spectrometer

María de Lourdes Gallardo Parrodi[1], José Luis Ruvalcaba Sil[2]
[1]Museo del Templo Mayor, Instituto Nacional de Antropologia e Historia, INAH. Seminario No. 8, Centro Histórico, Mexico DF 06060, Mexico.
e-mail: gallardoparrodi@gmail.com
[2]Instituto de Física, Universidad Nacional Autónoma de México UNAM. Apdo. Postal 20-364, México D. F. 01000, Mexico. e-mail: sil@fisica.unam.mx

ABSTRACT

In four offerings of the Great Temple of Tenochtitlan five groupings of *Pinctada Mazatlanica* shell pendants were found. Due to the burial conditions, damages on the surfaces can be observed in almost all the objects. In order to assess the deterioration degree, we used a visible light spectrometer. This is an inexpensive method to determine qualitatively the reflectance of the light at the surface that is directly related to the amount of organic material remains in these objects. This data may be used as a conservation marker for monitoring the collection and it can provide outstanding information to preserve the fragile shell pendants using a non-destructive method.

INTRODUCTION

The Great Temple Project was created in order to rescue and study the archaeological traces of the culture that dominated the Central High Plateau until the arrival of the Spaniards in the XVI century. Since 1978, more than 130 offerings have been recovered; all of them had different materials and were deposited for the Mexica's gods in the main building of their society: The Great Temple. The offerings were the mean by the Mexica communicated with their gods, because of that the offered objects have a lot of importance, since not only the deities, the represented forms and concepts are important, but the selected raw materials, its origins, manufacture and particular characteristics also reveal crucial data for the integral comprehension of this culture. The archaeological project has distinguished itself since its beginning for favoring a continuous interdisciplinary practice. By this way, the objects have been addressed in relation to its contexts, establishing their symbolic relations, their meaning and in many cases we have been able to understand better its behavior inside its immediate environment in the burial particular conditions. Moreover, some alterations suffered by the objects during its original use and after some treatments subsequent to the excavation have been identified in some cases. In the case of alterations of the pieces it has been possible to establish the cause, the mechanisms and the effects of deterioration. Considering that the archaeological collection now has more than ten thousand pieces and is still growing, this kind of approximations have been made in the most representative cases.

Among the archaeological materials from the Great Temple, we highlight the presence of shells (Figure 1), which were laid down in the offerings in natural or worked state. The amount of them that can be found is remarkable; this confirms the importance which they had in the pre-Hispanic period. The Mexica considered them as precious materials related to water and fertility, critical elements for life in this society. Also, they were distinctive attributes of gods, rulers and notable warriors, fact confirmed in some codex´s images, wall paintings and sculptures. According to the biological identifications, it is known that the shells laid down in the offerings are from numerous species that come from the Pacific, the Atlantic and the Caribbean. Its presence in archaeological contexts allows us to infer the work that took its gathering, transportation and eventual transformation into elaborated pieces, which obviously increased the value they had for the Mexica.

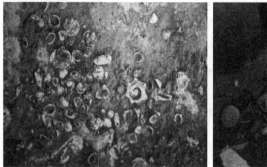

Figure 1. Shells and other artifacts discovered in the Great Temple offerings.

Of all these shells, we highlight the presence of one species: the *Pinctada mazatlanica*, a nacreous shell of silver color from which some of the most beautiful pieces of the archaeological collection were elaborated. It comes from the Panamic malacological province which distribution is from Baja California to the north of Peru. It is a coastal benthonic species that adheres to the firm substratum at the end of its larval life and its most important characteristic is the production of pearls. The inner part of the valves is bright white and pearly, unlike the external part which is usually darker, grey, golden or black.

The elaborated objects from this species were found mainly in the offerings of the fourth constructive stage of the Great Temple (1440-1469) that belonged to Moctezuma I (Figure 3). The number of these pieces makes up the 26.5% of the complete objects total from the collection [1]. However most of the objects made from this shell had alterations that were mainly originated during its burial, due to the particular conditions of the offering or by their interaction with other elements of the offerings.

PREVIOUS STUDIES

The study of the alterations of this kind of materials has been one of the tasks of the Conservation area of the Great Temple Project, since the huge amount of shell objects that have been recovered during the seven excavation seasons make its treatment an ongoing work. The deterioration and conservation of archaeological shell materials form the Great Temple has been approached previously, although in a punctual manner. There is a work that deals with its degradation and conservation [2], the study addresses the case of decay on the worked tinlkers of the *Oliva* genre and proposes certain treatments for consolidation, cleaning and reassembling of fragments. This study did not consider a diagnosis method which identifies the alteration degree of the pieces and can be implemented more easily to a considerable amount of them. The implementation of non-destructive analysis in the archaeological materials of the Great Temple collection for the study of their conservation has been carried out for the collections since 1999. Since then, several techniques for the identification of salts and constituents materials on wall painting have been successfully employed [3], for the recognition of the manufacturing techniques and degradation of copper bells [4], green stone objects, turquoise [5], textile and paper-like materials [6].

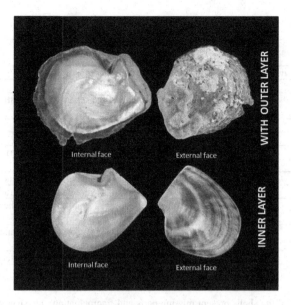

Figure 2. *Pinctada mazatlanica* nacreous shell.

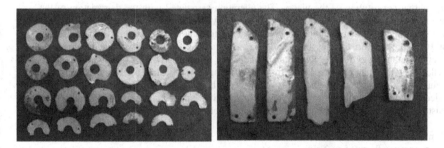

Figure 3. Some of the shells artifacts studied in this work.

SHELL OBJECTS DETERIORATION

Since the first observations made on shells at the beginning of the project, it was evident that the most frequent decay is the resistance loss due to the degradation of the constituent materials, particularly, the organic components like conchiolin. This loss can be noticed macroscopically in the pieces like a structural weakening and an opaque, porous and dusty surface. To carry out its preservation it was essential to diagnose and estimate the damage. After that, we developed register protocols which evaluate macroscopically the shells appearance and the degradation of the organic component through superficial aspect inferences, relating the variables of magnitude and frequency of the alterations presented on each object. The main identified alterations on the objects are described in Table 1.

Table 1. Deterioration observed in the shells pieces.

Alterations	Description
Loss of superficial layers	Partial lack of outer thin laminas
Missing	Total loss of constituent material
Fissure	Elongated opening product of the separation of constituent materials at superficial level
Detachments	Fragments separated from the constituent material
Cracks	Reticular pattern of cracks on the shell surface
Dusty	Superficial disaggregation as a result of protein matrix degradation
Abrasion	Surface erosion as a result of friction or erosion phenomena
Bright loss	Opacity of the surface due to the protein matrix degradation
Concretions	Surface deposition in the form of crust attached to the surface of a foreign material
Stains	Limited areas with distinct color and appearance than the general surface
Discoloration	Loss of color intensity in limited sections of the surface
Dirty	Dust and foreign materials deposited in the surface

Although many of the shells present missing parts and fragmentation –physical damages related with the burial context or with the direct fracture during original usage or its interaction with other elements in the offerings-, it is clear that this deterioration do not alter the internal constitution of the pieces. The alteration associated with the degradation of the organic part had not been possible to quantify directly or microscopically on the specimen so far. Nevertheless, this measurement is essential for making more accurate diagnoses for an assertive intervention on the shells collections. Because of that, we worked developing a specific methodology for the evaluation of the shells materials. Then, a study of five sets of shell pieces of *Pinctada mazatlanica* from the offerings 24, 88, Chamber II and Chamber III is currently on progress. We chose this set of pieces because it is representative of the entire collection conditions and the objects present highly differentiated states of preservation, from resistant, shiny, and iridescent complete pieces to much degraded elements, some of them like dust. All the pieces of these sets are laminated pendants with edges and holes. The pieces are of minor format, their size varies between 1 and 10 cm of maximum length and their thickness range from 0.5 to 2 mm.

PROPOSAL OF CHARACTERIZATION

Microscopic analysis was considered as an alternative to corroborate the macroscopic observations of the alterations. At first, Raman technique was used to study the calcite and aragonite contents of the shells as well as for the organic components, since this technique provides a specific spectrum for each material [7]. Measurements were made on both surfaces of the pieces with different parameters. When analyzing the results obtained with this procedure it was evident that due to the particular structure of the sample, -alternating juxtaposed layers of calcite and aragonite impregnated with conchiolin, which gives de pearly look to the *Pinctada mazatlanica*-, the surface was heterogeneous and in some regions too reflective. In most of the cases there was a very low Raman signal and high fluorescence and the spectra were not useful for the deterioration measurement.

For this reason, it was decided to perform the characterization of the conchiolin remains and the deterioration conditions of the shells by the use of an optic fiber spectrometer to measure the reflected light by the surface of the pieces. In this way the degree of surface alteration will determine the light dispersion. The device measures the scattered light in the range from 300 to 1000 nm.

METHODOLOGY

The first step consisted on the macroscopic review of the pieces from which three alteration degrees were established, preliminarily they are grouped into those with good, regular and bad conservation state. Since the pattern of deterioration observed was similar for the collection, it was decided to select a representative sample of 30 pieces, ten for each grade of alteration. The studied groups of shell pieces in various conservation conditions are shown in Figure 4.

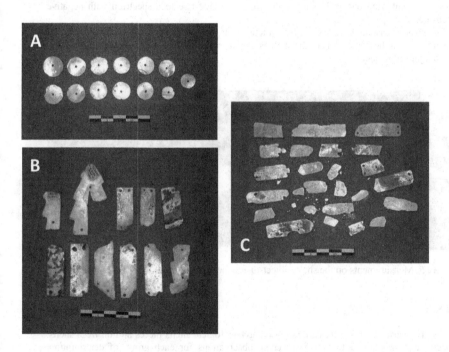

Figure 4. Groups of shell pieces with different conservation conditions. A) Good conservation conditions, low deterioration. B) Medium deterioration, C) Bad condition with higher deterioration.

The spectrometer was used in the conservation laboratory of the museum. Whereas some pieces had been previously submitted to various conservation treatments, it was decided to determine the parameters of analysis, carrying out measurements on new shells pieces from experimental archaeology tests and not on the archaeological ones.

The measurements were done in both faces of the selected 30 archaeological pieces in periods of 30 ms, with an Ocean Optics USB4000 spectrometer with a probe and optical fibers of 400 μm diameter for the 300 to 1000 nm range, averaging 20 spectra. A halogen source from Ocean Optics was also used (Figure 5). The measurements were carried out perpendicular to the surface to determine direct reflection. Shell spectra were compared to the opaque white color reference used for the spectrometer calibration (WS-1 diffuse reflectance standard). The measured spectra will correspond to the difference between the shell spectrum and the opaque white color reference spectrum. Thus, a material in good condition with high iridescence and high light reflection will produce a spectrum with positive intensities in the spectrum range. On the contrary, in those cases where a deficiency in the light reflection or light absorption is registered, it will indicate some alteration degree and shell structure decay. Low light scattering and reflectance reductions will give rise to a spectrum with negative intensities.

Control measurements were also practiced in non-archaeological shells, using the spectrometer in the white parts of the shell as well as in the darkest zones, to test the variation in the light reflectance.

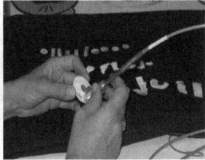

Figure 5. Measurements on the shells objects using the Ocean Optics USB4000 spectrometer.

RESULTS

Sixty measurements were carried out on archaeological shells pieces and on fresh shells. Spectra were compared to the macroscopic observations for each group of deterioration. Typical spectra of non-archaeological shell pieces are shown in Figure 5.

The spectra obtained on the fresh shell served as reference of the optimum conservation state parameter for being a material that has not been affected by the weathering processes of the burial. Also, the graph of the internal face of the fresh shell illustrates the spectrum of higher reflection. The white internal regions present almost three times more light reflection at the spectra maximum than the external darker face.

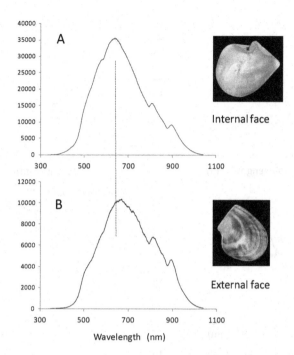

Figure 6. Measurements on experimental non-archaeological shell pieces A. Internal face white area. B. External face dark area.

Typical spectra obtained for the archaeological pieces of the three groups are shown in Figure 7. For the group of good conservation, in general the spectra have high positive intensities corresponding to a surface with high reflectance (Figure 7 A and B). Some peaks may be observed in the range of 600-650 nm and in some cases in the near infrared regions around 800 nm. The intensities are lower than the observed for the fresh shell (Figure 6).

For the medium conservation condition group the spectra are in most of the cases positive (Figure 7 C) but the higher intensities are between 1/3 to 1/4 of the average maximum intensity of the good conservation condition group. In some cases, the intensities are negative but with a low intensity (Figure 7 D). This means that the surface reflection for this group is similar to the white color reference but generally the surface is more reflective.

For the bad conservation condition group, the spectra present negative intensities (Figure 7E). This corresponds to an opaque surface with low reflectance and low scattering of the light. Nevertheless, in some regions the spectra have low positive intensities (Figure 7F). The light reflection is much lower than the white color reference but in some areas the spectra are similar to those observed for the medium conservation condition group.

In general the observed behavior in the spectra corresponds to the expected surface deterioration conditions and the measurements indicate an heterogeneous surface. The measured surface with the optical fiber is about 1 mm^2, making it sensitive to alterations which are not observed macroscopically. The microscopic measurements are then necessary to evaluate accurately the state of deterioration of the shells surface and the entire object.

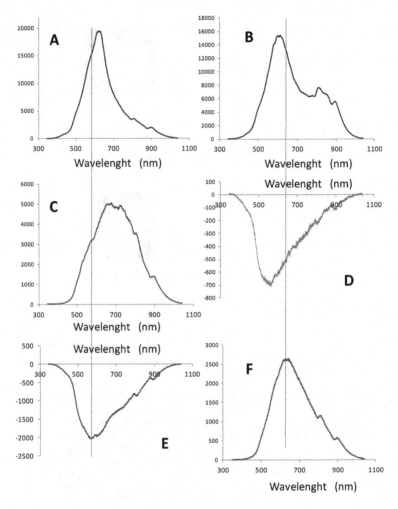

Figure 7. Typical spectra from archaeological shell pieces. A-B) Good conservation conditions, low deterioration. C-D) Medium deterioration, E-F) Bad condition with higher deterioration.

FINAL REMARKS

Our results allowed us to propose a fast, simple, inexpensive and portable methodology for archaeological shell collections diagnosis, and a new tool to study this kind of materials.

The macroscopic observation provides valuable data for the decay analysis as a first criterion, but it is necessary to complete the evaluation with microscopic measurements to confirm the macroscopic assessment. Since the decay patterns are generally consistent in the conchological collections is feasible to determine representative samples for the structural degradation analysis and to relate them with the surface observations. Accurate and fast results obtained with the spectrometer for alteration evaluation will mean more adequate procedures of conservation and more suitable materials for the archaeological objects treatment. Also, it is possible to adjust the measurement parameters to other shell species with different surface characteristics –like color and iridescence.

The proposed methodology can be completed and improved using references for other measurements, such as low and high-reflectivity specular reflectance standards. Finally, it is unavoidable to mention the importance of continuing the interdisciplinary work in order to increase the integral knowledge of the archaeological shells collections.

ACKNOWLEDGMENTS

Authors acknowledge the supports of UNAM PAPIIT IN403210 project as well as the CONACYT Mexico grant 131944 to carry out this study. This research has been performed in the frame of the Non Destructive Study of the Mexican Cultural Heritage ANDREAH network (www.fisica.unam.mx/andreah).

REFERENCES

1. A. Velázquez Castro, *La producción especializada de los objetos de concha del Templo Mayor de Tenochtitlan*. Instituto Nacional de Antroplogía e Historia INAH, Colección Científica 519, Mexico, 2007.
2. D. M. Grimaldi, (1997) *La colección de concha del género Oliva del Museo del Templo Mayor: un estudio para la conservación de concha arqueológica*. Dissertation on Conservation. Escuela Nacional de Conservacion, Restauración y Museografía, Instituto Nacional de Antroplogía e Historia INAH, Mexico 1997.
3. J. Miranda, M.L. Gallardo, D.M. Grimaldi, J.A. Román-Berrelleza, J.L. Ruvalcaba-Sil, M. A. Ontalba Salamanca, J.G. Morales, *Nuclear Instruments and Methods in Physics Research* B 150 (1999) 611-615.
4. N. Schulze, ¿Cobre para los dioses y oro para los españoles? Las propiedades sociales y simbólicas de un metal sin importancia, in: *Producción de bienes de prestigio ornamentales y votivos de la América antigua*, E. Melgar, E. Solís, R. González Licón (comp.), Syllaba Press., Mexico 2010, 75-85.
5. J.L. Ruvalcaba J.L, E. Melgar, Th. Calligaro. (2011), Manufacturing Analysis and Non Destructive Characterisation of Green Stone Objects from the Tenochtitlan Templo Mayor Museum, Mexico, in *Proceedings of the 37th International Symposium on Archaeometry, 13th - 16th* May 2008, Siena, Italy. Turbanti-Memmi ed., Springer, XLV, Heildelberg, 2011, 299-304.
6. M.L. Gallardo, M.L. La conservación de las ofrendas de la Casa de las Ajaracas y de la Casa de las Campanas, in: *Arqueología e historia en el Centro de México. Homenaje a Eduardo Matos Moctezuma*. L. López, D. Carrasco, L. Cué (comp.) Instituto Nacional de Antropología e Historia INAH, Mexico 2006, 555-567.
7. F. Agulló-Rueda, Espectroscopía Raman, in: *La Ciencia y el Arte. Ciencias experimentales y conservación del patrimonio histórico*. Vol. I, M. del Egido, T. Calderon (coords.) Instituto del Patrimonio Histórico Español. Instituto de Cultura, Madrid, 2008, 117-124.

Mater. Res. Soc. Symp. Proc. Vol. 1374 © 2012 Materials Research Society
DOI: 10.1557/opl.2012.1388

Identification of Microorganisms Associated to the Biodegradation of Historic Masonry Structure in San Francisco de Campeche City, México

Rocío G. Escamilla Pérez[1], Javier Reyes Trujeque[1], Tezozomoc Pérez López[1], Víctor Monteón Padilla[2], Ruth López Alcántara[2]

[1] Centro de Investigación en Corrosión, Universidad Autónoma de Campeche. Avda. Agustín Melgar s/n entre Juan de la Barrera y Calle 20. Colonia Lindavista. San Francisco de Campeche, Campeche, México. e-mail: javreyes@uacam.mx

[2] Centro de Investigaciones Biomédicas. Universidad Autónoma de Campeche, Campeche Mexico.

ABSTRACT

Tropical climate create ideal conditions for the development of microbial communities associated with biodegradation of historic buildings made with stony materials. This is the case of Fort San Carlos, a historic colonial building representative of military tendencies during the XVII century in San Francisco de Campeche City. In this study the Polymerase Chain Reaction (PCR), was used to identify microorganisms related with the biodegradation of its masonry structure. Specific primers for amplification of 16S and 18S ribosomal RNA genes were used for organisms identification by PCR. Amplification products were sequenced and after that compared with GENBANK nucleotide database using-BLASTn. Results indicated that microbial communities associated to biodegradation of the Fort San Carlos are bacteria from the Phyla Cyanobacteria, Proteobacteria and Actinobacteria.

INTRODUCTION

Nowadays, there is a special interest to understand biological and non-biological processes involved in biodegradation of ancient buildings, especially those constructed with calcareous materials. In tropical climates, it is favored by a high water availability which facilitates the development of microbial. Aesthetical and chromatic changes in buildings surface are visual consequences of microbial activity. Indeed metabolic activity of microbial community can act synergistically with environmental factors to produce physical and chemical changes inside the stony structures, sensitizing the materials and compromising their stability.

Modern molecular biology procedures allow rapid, efficient identification of members in microbial biofilms associated to several solid phases, including industrial pipes, bioreactors, artificial reefs, and ancient monuments.

Actual research tendencies concerning to the biodegradation of historic buildings are focused in the analysis of biofilms, the development of novel molecular techniques to identified associated microorganisms and the use of biotechnology to generate new conservation procedures [1,2].

San Francisco de Campeche City, is located in the occidental coast of the Peninsula of Yucatan. It has about 1500 civil, military and religious buildings builts between XVI to XVIII centuries. In 1999, colonial buildings inside the Spanish fortified walls was included into the UNESCO's Cultural Heritage List.

Fort San Carlos is a military structure representative of the defensive system constructed around the Spanish village by colonial authorities. The Fort crowned the southeastern corners of the irregular hexagonally-shaped bastion-and- rampart fortified system surrounded San Francisco de Campeche´s urban core. This defensive public-works project was designed to protect both local governmental offices and Spanish colonial residences from continuous pirate attacks during XVI to XVIII centuries. Local calcareous materials including quarried limestone blocks, and sand, slake lime and carbonate clay marls (known as sahacab in Yucatán), combines to erect the fort, forming a solid masonry structure. Now a day, the structure shows several symptoms of degradation consequence of their long interaction with the characteristic tropical environment [3, 4].

Previous works report the use of FTIR, SEM/EDX and XRD to analyze the mineral alteration of stony materials from the walls of the building [3-5]. In this study, several components of limestone and traditional mortars employed during the construction of the Fort and posterior preservation works were identified. Authors also report the neoformation of calcium carbonates ($CaCO_3$), bassanite ($CaSO_4.1/2H_2O$) and the calcium oxalates: whewellite ($C_2CaO_4.H_2O$) and whedellite ($C_2CaO_4.2H_2O$). Formation of calcium oxalates was consequence of limestone dissolution by oxalic acid excretion from microorganisms like cyanobacteria and lichens [6-8].

In the present study, Polymerase Chain Reaction (PCR), was used to identify microorganisms related with the biodegradation of the masonry structure of the Fort. Specific primers for amplification of 16S and 18S ribosomal RNA genes were used for organisms identification by PCR. Amplification products were sequenced and compared with GENBANK nucleotide database using BLASTn algorithm.

MATERIALS AND METHODS

The Building

Fort San Carlos is an irregular pentagonally-shaped masonry structure located at the 8th street, just in front of the principal Government office buildings and the State Congressional offices of Campeche (Figure 1). Its construction dates back from 1680. It has a trapezoidal plant structure reinforced by buttress at the south façade. Its walls are topped by a perimeter cornice and width battlements. Three watchtowers are observed in the angles of the building giving the Golf of México shoreline. All the construction was built by using calcareous materials from the region that include limestone quarry blocks, slake lime and sascab [3].

Sampling

Figure 2 presents the sampling layout used during the study. Eleven samples were collected by using a sterilized spatula, kept in sterile conic tubes and storage at - 40° C (Revco Ultrafreezer), until molecular analysis was carried out.

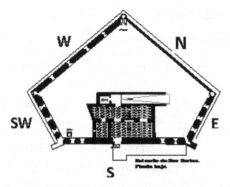

Figure 1. Pentagonally structure of Fort San Carlos and location of their walls: North (N), South (S), East (E), Southwest (SW) and West (W).

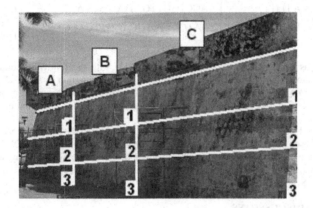

Figure 2. Sampling performed in one side of the San Carlos Bulwark.

DNA extraction

DNA was extracted and purified using phenol:chloroform:isoamyl alcohol technique (25:24:1) [9]. Saccharomyces cerevisiae (eukaryotic) and *Escherichia coli* (prokaryote) were used as controls. The presence of DNA was confirmed through electrophoresis in agarose gel (2%) at 85 V/, and reveled with ethidium bromide (1µg/ml). The concentration and purity of DNA was evaluated using UV-Visible spectrophometric method (Jenway-6305). Measure of DNA dissolution (1:200, in sterile water) at 260 and 280 nm in order to estimated the DNA/protein rate.

PCR amplification of 16S and 18S rDNA

All reactions were carried out in final volume of 50 µl containing: 1 µl of genomic DNA (200 ng), 1 µl of dNTPs (50 – 300 ng according the specific primer), 5 µl of 10X PCR buffer, 1.5 µl of MgCl$_2$ (50Mm) and 1 µl of Taq DNA polymerase (2.5 U). Primers used in the study are shown on table 1.

Table1. Selected primers used for the amplification of samples and controls.

Primers	Sequences
616F [10]	AGAGTTTGATYMTGGTGGCTCAG
907R [11]	CCGTCAATTCCTTTGAGTT
Cya 106 F [12]	CGGACGGGTGAGTAACGCGTCGTGA
Cya 781Ra [13]	GACTACTGGGGTATCTAATCCCATT
Cya 781Rb [13]	GACTACAGGGGTATCTAATCCCTTT
Euk A [14]	AACCTGGTTGATCCTGCCAGT
Euk B [15]	TGATCCTTCTGCAGGTTCACCTAC

The PCR was performed in a Thermocycler (TC-512) with the following conditions: 2 minutes denaturation at 95 °C, followed by 35 cycles of 15 s denaturation at 95 °C, 55 °C of 15 s annealing and 1min extention at 72 °C. A final extention step of 72 °C during 10 minutes. PCR products were verified through electrophoresis in agarose gel (2%) and ethidium bromide (1µg/ml) staining.

For PCR products purification and sequencing, the Wizard ADN Clean-Up System from (Promega) was used, following the protocol recommended by the fabricant. DNA samples amplification were sequenced at the UNAM Institute of Biotechnology, employing primers concentration at 10 pmol/µl. The DNA sequence comparison was accomplished by the Basic Local Alignment Search Tool (BLAST).

RESULTS AND DISCUSSIONS

Fort San Carlos shows several signs of deterioration, including the development of microbial communities characterized by abundant black color strains. White, green and light yellow colonies also can be observed although in smaller abundance. It is important to know the preferential microbial development in free mortar and paint areas of the building. In this order the E wall showed the most microorganisms affected area of the Fort (Figure 3).

Figure 3 shows the integrity analysis of the 11 samples and controls by Agarose gel electrophoresis, noticing the presence of stripes corresponding to high molecular weight DNA. On the other hand, Figure 4 shows the PCR products of the samples M1, M2 and controls. The PCR analysis allowed obtaining products at ~500 pb, ~1000 pb, ~650 pb and ~2750 pb.

Figure 3. E wall of the Fort San Carlos.

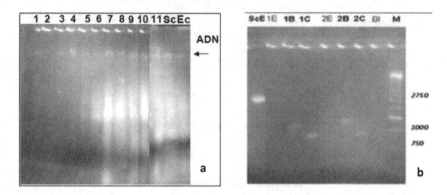

Figure 4. (a) Integrity analysis of samples and controls. (b) Profiles obtained using the primers, E= Eukaryote, B= Bacteria, C= Cyanobacterias, Bl= Blanco, M= Marker y Sc= *S. cerevisiae.*

The sequenced PCR products obtained from the samples M1B, M2B, M4B, M6B, M8B, M10B, M11B and M11C, belonged to those amplified bacteria and cyanobacteria primers. Table 2 shows the identified microorganisms taxonomic affiliation, percentage of similarity and number of bases, according to the GenBank (NCBI).

Table 2. Taxonomic affiliation of the microorganisms identified at Fort San Carlos.

# of access	Samples	Microorganisms	% similarity
AY972169.1		*Pseudomona plecoglossicida* P13 16S rRNA	97
AB461012.1	4B	Pseudomonas sp. IK1_29 16S rRNA gene	97
EF682071.1		*Pseudomona putida* DN1.2 16S rRNA gene	97
EU073968.1		Brevibacterium CMGS4 16S rRNA gene	97
GU296674.1		*Pseudomona aeruginosa* ANSC 16S rRNA	99
GU183597.1	2B	**Uncultivable** *bacteria* NMG51 clone 16S rRNA gene	99
AY972169.1		*Pseudomonas plecoglossicida* strain P13 16S rRNA gene	97
EF071544.1		**Uncultivable** *bacteria* JordanB10 clone 16S rRNA gene	83
FJ230829.1	6B	**Uncultivable** *bacteria* Prehnite45 clone 16S rRNA gene	89
DQ914865.2		*Chroococcidiopsis sp.* CC3 16S rRNA gene	83
AM934704.1		*Pseudomona sp.* Incultivable clon 16S rARN	97
AY699600.1	8B	Bacteria de suelo incultivable SAL2d22 clone 16S rRNA gene	97
AY972169.1		*Pseudomona plecoglossicida* rRNA 16S gene	97
FJ230797.1		**Uncultivable** *bacteria* QuartzC3 clone 16S rRNA gene	83
FJ589716.1	10B	*Chroccocales* cyanobacteria LEGE 060123 16S rRNA gene	84
AJ344552.1		Chroococcidiopsis sp. BB79.2 16S rRNA gene, SAG 2023	82
AM746687.1		**Uncultivable** *bacteria* parcial 16S rRNA gene, clone A9	95
EF032781.1	11B	**Uncultivable** *cyanobacteria* HAVOmat11 clone 16S rRNA gene	93
FJ790611.1	11C	**Uncultivable** *bacteria* QB35 clone 16S rRNA gene	86

Bacteria group, was characterized by the presence of *Proteobacteria* (*Pseudomona sp.*, 97 % similarity) in samples M4, M8 and M2, and A*ctinobacteria* (*Brevibacterium sp.*, 97 % similarity*)* in sample M4. These microorganisms have been related with the biodegradation of the stone buildings [16]. Some kind of sulfur-nitrifying bacteria like *Bacillus sp.*, *Micrococcus sp.* or *Pseudomonas sp.*, are common in stone surfaces [17, 18].

They have been reported as the major colonizers in Mayan monuments at the archeological site of Uxmal, México (Ortega–Morales 2004). It is known that *Pseudomonas* sp. produce mucilaginous biofilms that cause biodeterioration and chemical dissolution of stone minerals [18, 20]. It is important to note that in all the samples (except sample M4B); uncultivable bacterias ranging from 97 to 83 % similarity were reported.

On the other hand, cyanobacteria group showed the presence of *Chroccocales* (sample M10) and *Chroococcidiopsis* (sample M6) with 82 and 83% of similarity respectively. Samples M6, M11B and M11C reported the presence of uncultivable cyanobacteria ranging from 83 to 93% similarity. *Chroococcidiopsis* cyanobacteria are considered cosmopolites and cryptoendolithic organisms [17]. These microorganisms have been observed forming the biomass of Brazilian monuments with higher degree of deterioration [21].

Otherwise, Eukaryotic group are common organisms widely distributed in stone surfaces and monuments around the world [20, 22, 23]. Nevertheless, the methodology employed during this study do not permitted to obtain PCR products from this group. Otherwise, unpublished optical and SEM micro-images indicated the presence of abundant mycelium entrapped into the structure of the calcareous materials samples from Fort San Carlos [24].

CONCLUSIONS

Biodeterioration related species were identified in historic buildings edificated with stone materials. Bacteria belonging to Cyanobacteria, Proteobacteria and Actinobacteria phyla were the most dominant in this study.

Methodology used did not report the right information about eukaryote, in this order studies about that phyla most be carried out. However, the results showed that the adaptation of molecular biology techniques is of a big importance in the identification of microorganisms related to the biodeterioration of historic buildings, such as Fort San Carlos.

Complementary studies must be performed with the aim of providing the right information about the microbial flora existing in historical buildings, so the importance of biodeterioration processes can be determined.

ACKNOWLEDGMENTS

This contribution was possible thanks to the support of FOMIX CAMP2005-C01-025 Project. Also thanks to Centro INAH-Campeche for their giving facilities to the development of the project.

REFERENCES

1. D. Allsopp, K. Seal C. Gaylarde, *Introduction to Biodeterioration*, Cambridge University Press, Cambrige UK, 2004, 179-190.
2. J. Gonzalez, Overview on existing biomolecular techniques with potential interest in cultural heritage, in: *Molecular Biology and Cultural Heritage*, C. Saiz-Jimenez ed., Balkema, Lisse, 2003, 3-13.
3. J. Reyes, G. Gutiérrez, G. Centeno, D. Treviño, P. Bartolo, P. Quintana, J. Azamar and T. Pérez, Proceedings 1th Historical Mortar Conference, Portugal, 2008.

4. J. Reyes, F. Torres, M. Miss, F. Corvo, H. Bravo, P. Bartolo-Pérez and J. Azamar- Barrios, in *Environmental degradation of infrastructure and cultural heritage in coastal tropical climate*, J. González-Sánchez, F. Corvo and N. Acuña-González (eds.), 2009, 115-142.
5. F. Corvo, J. Reyes, C. Valdes, F. Villaseñor, O. Cuesta, D. Aguilar and P. Quintana. Water, *Air, Soil Pollut.* 205 (2010) 359-375.
6. M. Del Monte, C. Sabbioni. *Sci. Tot. Environ.* 50 (1986) 165-182
7. I. Rampazzi, A. Andreotti, I. Bonaduce, M. Colombini, C. Colombo, L. Toniolo. *Talanta* 63 (2004) 967-977.
8. J. Arocena, T. Siddique, R. & S. Thring. *Dev. Soil Sci.* 70 (2007) 356-365.
9. F. Ausubel, R. Brent and R. Kingston, Curr. Protoc. Mol. Biol. Short Protocols in molecular biology, vol. 2, 2002.
10. J. Zimmermann, J. Gonzalez, C. Sáiz-Jiménez, *Geomicrob.* 22 (2005) 379–388.
11. W. Weisburg, S. Barns, D. Pelletier, D. Lane, *J. Bacteriol.* 173 (1991) 697-703.
12. U. Nübel, F. Garcia-Pichel, G. Muyzer. *Appl. Environ. Microbiol.* 63 (1997) 3327–3332.
13. A. Bonazza, C. Sabbioni, N. Ghedini, B. Hermosín, V. Jurado, M. J. González and C. Sáiz-Jiménez. *Environ. Sci. Technol.* 41 (2007) 2378-2386.
14. B. Diez, C. Pedro´s-Alio, T. Marsh, R. Massana. *Appl. Environ. Microbiol.* 67 (2001) 2942–2295.
15. J. González and C. Saiz-Jiménez. *Int. Microb.* 8 (2005) 189-194.
16. A. Gorbushina. *Environ. Microbiol.* 9 (2007) 1613–1163.
17. A. Gorbushina, N. Lyalikova N. D. Vlaso and T. Khizhnyak. Microbiol. 71 (2002) 350–356.
18. B. Ortega-Morales, G. Hernández-Duque, *Ciencia y Desarrollo* 24 (1998) 48-53.
19. B. Ortega-Morales. *Curr. Microbiol.* 40 (2004) 81-85.
20. H. Videla, P. Guiamet, S. Gómez de Saravia, in: *Biodeterioro de monumentos históricos de Iberoamérica*, H. Videla ed., CYTED, Sevilla, 2000, 5-17.
21. C. Crispim, P. Gaylarde, C. Gaylarde. *Int. Biodeter. Biodegrad.* 54 (2004) 121-124.
22. G. Gómez-Alarcón, B. Cilleros, M. Flores, J. Lorenzo, *Sci. Tot. Environ.* 167 (1995) 231-239.
23. P. Gaylarde, C. Gaylarde. *Int. Biodeterior. Biodegrad.* 55 (2005) 131-139.
24. R. Escamilla-Pérez. Undergrade Thesis. College of Chemistry and Biological Sciences. Universidad Autonoma de Campeche, Campeche, 2010.

Mater. Res. Soc. Symp. Proc. Vol. 1374 © 2012 Materials Research Society
DOI: 10.1557/opl.2012.1389

Chemical Evolution of the Volcanic Tuff from the Santa Mónica Church in Guadalajara, Mexico

Nora A. Pérez and Enrique Lima
Instituto de Investigaciones en Materiales, Universidad Nacional Autónoma de México.
Circuito exterior s/n, Ciudad Universitaria, Delegación Coyoacán, CP 04510, México D.F.
e-mail: norari.perez@gmail.com , lima@iim.unam.mx

ABSTRACT

The Santa Mónica Church is one of the most representative buildings in Guadalajara, Mexico as it is the finest Solomonic Baroque temple in the city. The church was built in the XVIII century with different types of volcanic tuffs, which have been studied from the macroscopic level to the structural level with the aim to determine the deterioration degree of the church's tuffs.

Textural, morphological and structural properties of Tuff were characterized using X-ray powder diffraction (XRD), infrared spectroscopy (FT-IR ATR) and ^{29}Si and ^{27}Al magic angle spinning nuclear magnetic resonance (MAS NMR), nitrogen adsorption-desorption techniques, scanning electron microscopy (SEM), thermo-gravimetric analysis (TGA), compressive strength tests were also performed.

Characterization data has provided a comprehensive view of the alterations on the volcanic tuff of Santa Mónica Church. Then the study focused on proposing the best strategy for the understanding and conservation of Churches and other buildings in Guadalajara which have been built with the same stone. Currently, siliceous materials doped with aluminum are being tested as consolidate.

INTRODUCTION

The Santa Mónica church is located in the downtown of Guadalajara, Jalisco, Mexico (see figure1). It was built by Father Feliciano Pimentel between 1719 and 1720 and was part of the Augustinian Nuns convent until it was confiscated by the government during the Reform war. Through a public auction, the lawyer Dionisio Rodriguez bought the convent and gave it to the Diocesan Seminary of Guadalajara, whose building had also been confiscated by the government.

Later, the diocesan seminary purchased the convent, and demolished it to build an *art nouveau* type building, designed by French architects in the early twentieth century. However, before the Cristero War, the governor Guadalupe Zuno confiscated the building again and the convent was given to the Mexican Army, which established the XV Military Zone which operates there until now [1].

The church is built with two kinds of volcanic tuff: the Yellow tuff and the White tuff; this type of rock is important since it is the main building material for most buildings in Guadalajara's downtown. The types of degradation that are present are salt efflorescence, erosion, exfoliation, biological deterioration and also thermal expansion at the top of the building and anthropological factors such as loss of material due to collisions of the public transport.

The goal of the present study was to characterize both types of tuff in order to determine the alteration process of the rock. The results should be enough to propose a consolidating material compatible with the original rock.

Figure 1. Location of the Santa Mónica Church in the downtown of Guadalajara, Jalisco (right). Facade of the church (left).

EXPERIMENTAL

Samples

During the restoration procedures on the church in 2006, different rocks that showed advanced deterioration were removed and replaced with new tuffs, which were obtained from the original quarry used for the building of the Santa Mónica Church. For this study samples were taken of both white and yellow tuffs of the altered and fresh rocks.

Structural characterization

Powder X-ray diffraction (XRD) was performed with a Bruker-axs D8-advance diffractometer coupled to an X-ray diffraction copper anode tube. The X-ray diffraction patterns were recorded with a scintillation counter. A nickel filter selected the CuKα radiation. The identification of the minerals was performed conventionally comparing with the JCPDS files. The samples were characterized by Infrared Spectroscopy with attenuated total reflectance (FTIR-ATR). Spectra were recorded on a Nicolet 6700 Thermo Electron Corp spectrophotometer within the range 4000 a 400 cm^{-1} at room temperature with a resolution of 4 cm^{-1} . Thermogravimetric analyses were performed with TA Instruments equipment. Samples were treated at a heating rate of 10 °C min^{-1} from room temperature to 800 °C in nitrogen atmosphere.

Solid-state NMR spectra were obtained under MAS conditions using an ASX 300 Bruker spectrometer with a magnetic field strength of 7.05 T, corresponding to a ^{27}Al Larmor frequency of 78.3 MHz. A single pulse method was used with a π/2 pulse of 2 μs. The spinning frequency was 10 kHz .The ^{27}Al chemical shift was referenced using an aqueous solution of Al (NO₃)₃ as external standard. ^{29}Si MAS NMR spectra were obtained operating the spectrometer at a resonance frequency of 59.59 MHz with a High power decoupling pulse program (HPDEC). The spinning frequency was 5 kHz, and tetramethylsilane (TMS) was used as a reference.

Textural and morphological characterization

An Olympus trinocular petrographic microscope equipped with a camera and software for image analysis was used to determine petrographic composition, texture, and alterations. The morphology of the samples was established with a Leica-Cambridge Stereoscan 440 scanning electron microscope. Surface area analyses were performed on a BELSOPR-minill apparatus. The N₂ adsorption-desorption isotherms were determined at 77K by volumetric adsorption. Before the N₂ adsorption process, all samples were outgassed at room temperature for 16h. Surface areas were calculated with the BET equation and pore diameter values with the BJH method.

Macroscopic properties evaluation

Compressive strength of the rocks was measured according to the ASTM standard D2938-95. An Ocean Optics USB4000 spectrometer with a tungsten halogen light source (360-2500nm) was used to measure the colorimetric changes in the samples, the CIELab color space was employed.

RESULTS AND DISCUSSION

In all samples, either fresh or altered rocks, two phases were identified by XRD as can be seen in figure 2, sanidine ($K_{0.65}Na_{0.35}AlSi_3O_8$) which is an alkaline mineral feldspar and low trydimite (SiO_2). These compounds are generally present in volcanic rocks. An increase in the amorphous phase of the altered samples was observed between 10-20 degrees which is attributed to the devitrification process.

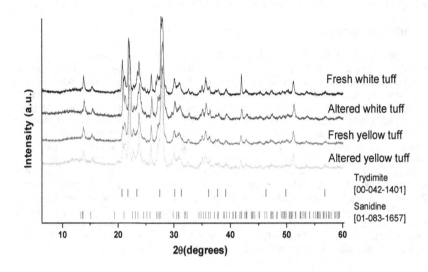

Figure 2. XRD pattern of the fresh and altered tuff.

These phases were also observed in the petrographic studies where the silicon dioxide matrix is present with crystals of sanidine (see figure 3) and other alkali feldspar, also lithic materials were present which are characteristic of volcanic explosions. The increase of iron oxides in the altered tuff was evident, which is in agreement with the colorimetric measurements. The red component (a*) of the altered white tuff increased 0.6 units while for the altered yellow tuff the increase was of 2.5 units. Also the saturation (C*) increases 7.8 units in the altered white tuff and 3.4 units in the altered yellow tuff.

FTIR spectra are consistent with the XRD results for the proportion of Si/Al, according to the frequency of the main band and its shoulder comparing to Breck [2] the relation of Si/Al is of 2.8. It is also possible to observe a change in the intensities of the bands associated with the loss of material in the damaged stones.

Figure 3. Petrography images of the fresh yellow tuff (left) and altered yellow tuff (right).

Table 1. FTIR Results of rock samples. s=strong, m=medium, mw=medium weak, w=weak, vw=very weak, sh=shoulder, b=broad, T=Si or Al.

Assignment	Wavenumber (cm^{-1})			
	Fresh White tuff	Altered White tuff	Fresh Yellow tuff	Altered Yellow tuff
Pore opening[4]		402 sb		
Pore opening[4]	416s,464vwsh		414 s	413s,462vwsh
Characteristic of the material[5]	539 m	54 m	540 m	537 m
Characteristic of the material[5]	578 mb	578 mb	581 mb	574 mwb
Characteristic of the material[5]	640 w,668w	639w,668w	639 w	639 w
Sym. Elongation internal TO$_4$ [4]	719 vw	718 vw	718 vw	
Sym. Elongation between TO$_4$ [4]	784 vw	784 vw	786 w	787vw,910vw sh
Asym. Elongation internal TO$_4$ [4]	1000 sb	994 sb	998 sb	1002 sb
Asym. Elongation between TO$_4$ [4]	1105 w	1100 w	1104 w	1095 vw

According to the frequency of the main asymmetric band there is an atomic fraction of 0.4 aluminum atoms in tetrahedral sites [3]. The band at 719cm^{-1} is present the spectra of all samples except in the altered yellow tuff which indicates that there is a loss of tetrahedral silicon, this is confirmed by the ^{29}Si NMR results.

The ^{29}Si NMR spectra (see figure 4) shows a broad signal at -99ppm for all samples, this signal can be assigned to Q^1, Q^2 or Q^3 units [4] present in silicate minerals. The signal at -112 ppm corresponds to tetrahedral silicon (Q^4) belonging to trydimite. The ratios between both signals were calculated from integration of corresponding peaks. The values obtained are for the fresh white tuff Q^4/Q^2=0.67, for the altered white tuff Q^4/Q^2 =0.92, for the fresh yellow tuff Q^4/Q^2=0.56 and for the altered yellow tuff Q^4/Q^2 = 0.53. In the white tuff there is an increase in Q^4 units for the altered rock in accordance to the devitrifying process of the rock; while for the yellow tuff there is a slight decrease in this signal which can be attributed to the presence of silanol species at the chemical shift or -101 ppm [5].

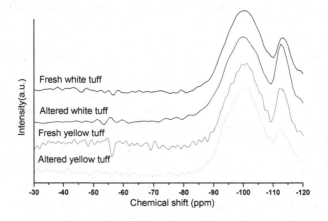

Figure 4. ^{29}Si NMR spectra for the volcanic tuffs.

From the SEM images (see figure 5) the increase in porosity and loss of material in the altered rocks is evident. This property was measured by the nitrogen adsorption technique, all the nitrogen adsorption-desorption isotherms are type IV and have hysteresis loops type H3 according to the IUPAC classification, which indicates a mesoporous material possessing rigid aggregates laminar type in long, thin pores. The specific surface area was calculated by the BET method, while the pore size distribution was estimated by the BJH method. There is a decrease in surface area of 80% for the white tuff and 18% in the yellow tuff compared to the fresh tuffs.

Table 2. Analysis results of the adsorption-desorption isotherms calculated with the BET and BJH method.

	Specific Surface Area BET (m²/g)	Average Pore Radius BJH (nm)
Fresh white tuff	5.77	1.88
Altered white tuff	1.09	2.41
Fresh yellow tuff	4.94	1.88
Altered yellow tuff	3.43	10.64

The average pore size of the fresh yellow and white tuff is the same; the altered white tuff increased 1.3 times the pore size while the altered yellow tuff increased five times the pore size.

The increase in porosity is confirmed by the TG analysis: the fresh rocks lost about one percent of their weight in the process of evaporation, while the altered white tuff lost 3.5% and the yellow tuff lost 5.5%. In the thermogram of the altered yellow rock there is a weight loss above 400°C which is associated with coordination water molecules with the aluminum surface (see figure 6). This was verified by ^{27}Al NMR. A signal at 0ppm indicates the presence of six coordinated aluminum and after thermal treatment of this sample the signal faded out (figure 7).

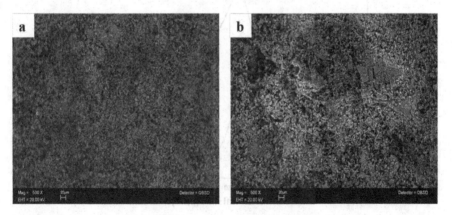

Figure 5. Scanning electron microscopy image of the (a) fresh white tuff, (b) altered white tuff.

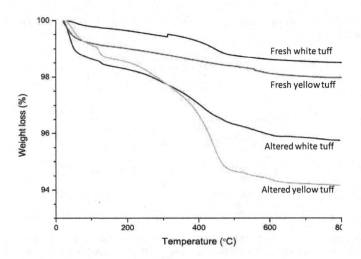

Figure 6. Thermograms of rock samples.

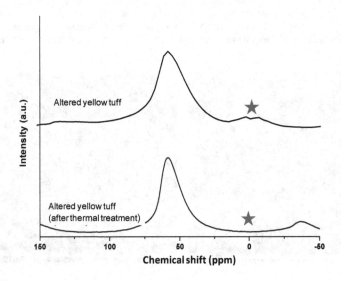

Figure 7. ^{27}Al NMR spectra showing the absence of the 0 ppm signal in thermally treated sample of the yellow tuff.

Finally in the compressive strength tests the fresh rocks have a larger load capacity than the altered stones as expected. The fresh white stone has an average maximum stress of 25MPa and the altered of 18MPa. The fresh yellow tuff has an average maximum stress of 40MPa and the altered of 11MPa.

CONCLUSIONS

From the different analysis it is concluded that the yellow tuff is more susceptible to alteration processes in comparison to the white tuff. A proposed degradation process involves the presence of aluminum atoms which increases the degree of hydration of the rock surface, promoting the hydrolysis of the silicon in the framework causing condensation reactions. The level of degradation is higher in the yellow tuff, probably because the reactions may be catalyzed by the presence of iron atoms that are initially in a higher amount in this type of tuff. Therefore the proposed consolidant must interact with the hydrolized aluminosilicate structure in order to restore the rock. According to the NMR results the most compatible material for the white tuff would be an aluminosilicate since the aluminum is extracted. For the yellow tuff, the most compatible material would be a silicate in order to equilibrate the Al-Si composition. These proposed materials are currently developed.

ACKNOWLEDGMENTS

Thanks to CONACyT for the M.Sc. scholarship and to Proyecto Templo de Santa Mónica for providing the rocks for this research. We thank Omar Novelo, Esteban Fregoso, Marco Vera and Miguel Canseco for the technical work in SEM, TGA, NMR and FTIR analyses respectively.

REFERENCES

1. Hallazgo en Templo de Santa Mónica,
   http://www.informador.com.mx/jalisco/2010/176860/6/hallazgo-en-templo-de-santa-monica.htm.
2. D. Breck, *Zeolite Molecular Sieves: Structure, Chemistry and Use,* Wiley Interscience Publisher, 1974, 416-418.
3. Sanidine R060313, The RRUFF™ Project Database.
   http://rruff.info/sanidine/display=default/R060313.
4. D. Breck, *Zeolite Molecular Sieves: Structure, Chemistry and Use,* Wiley Interscience Publisher, 1974, 420.
5. K.J.D Mackenzie,M.E Smith, *Multinuclear Solid State NMR of Inorganic Materials,* Elsevier Science, Oxford, 2002, 207.
6. X.S. Zhao, G.Q. Lu, A.K.Whittaker, G. J. Millar, H.Y. Zhu, *J.Phys.Chem.* B 101 (1997) 652.

Mater. Res. Soc. Symp. Proc. Vol. 1374 © 2012 Materials Research Society
DOI: 10.1557/opl.2012.1390

Correlation of Atmospheric Dust and Rainfall as Basalt Chemical Weathering Precursors

M. Teutli León[1], L.A. Térrez Tufiño[1], G. Jiménez Suárez[1], E. León Hernández[2], L.M. Tenorio Téllez[2]

[1] Facultad de Ingeniería, Benemérita Universidad Autónoma de Puebla, Puebla, Mexico.
e-mail: teutli23@hotmail.com

[2] Facultad de Arquitectura, Benemérita Universidad Autónoma de Puebla, Ciudad Universitaria, Puebla, Puebla, CP 72570, Mexico.

ABSTRACT

Basalt chemical weathering can be related to chemical composition of both atmospheric dust and rainfall, in published works authors have shown that weathered basalt exhibits a raise in anions like sulfate, nitrate, phosphate, chloride and carbonate, and doing a follow up of rainfall chemistry during 2009 at downtown Puebla, it was shown that rainfall has not an acid pH because there is a strong contribution from atmospheric dust.

In this paper it is reported obtained results for dust samples collected during the dry season, collection was done in 3 places exposed to different environment, analysis of solid samples clearly reflect anthropogenic activities since the highest oil and grease content corresponds to a site with high population, also mineral carbonate amounts 30% in weight. An analytical sample was prepared and the filtrate used to measure pH which ranges from 6.3 to 7.84, and conductivity from 11.91-13.87 mS-cm^{-1}. Main soluble ions range are as follow: sulfate 3.4-5.9 mg g^{-1}, nitrate 0.19-0.54 mg g^{-1}, chloride 0.7-8.91 mg g^{-1}, sulfide 4-7 mg g^{-1}, carbonate 304-364 mg g^{-1}, this last correspond to 10% of the mineral content. Also, metals were determined as total (applying an acid digestion) and water soluble, obtained results allow to affirm that there are highly soluble metals like Ca, Pb (up to 97%), moderately soluble ones such as Cu and Mn (60% and 20% respectively) and non soluble ones like Al, Fe and Zn.

It was confirmed that atmospheric dust has all properties, which could produce an alkalinization of rainwater. Also, its water soluble ionic content can be a source for those ions causing basalt weathering.

INTRODUCTION

Basalt chemical weathering at historical buildings is consequence of environmental quality at each place where buildings are located. Stone structural modifications can be related to chemical composition of both atmospheric dust and rainfall, in published works authors have shown that weathered basalt exhibits a raise in anions like sulfate, nitrate, phosphate, chloride and carbonate [1], and doing a follow up of rainfall chemistry during 2009 at downtown Puebla, it was shown that rainfall has not an acid pH because there is a strong contribution from atmospheric dust, authors arrived to this conclusion based on rainfall statistical correlations ions present in rainwater can have a common origin if they correlate positively as example in this published work marine components like sodium and chloride correlate with a 0.6 factor, otherwise sodium and carbonate correlate with -0.02 factor; from the same report calcium, magnesium, carbonate and sulfate should have a common origin which is different from rainwater since the correlation between CO_3 and Ca is 0.83, while CO$_3$ and Mg is 0.98; therefore, atmospheric dust should have alkaline properties [2]. Also, there is an assumption about how downtown commercial activities could exert some influence onto zonal pollution and modify dust chemical properties, since dust can act as nucleus to trap emitted pollutants.

Urban areas with a high number of vehicles circulating by, usually lead to atmospheric pollution problems in which fuel combustion produce fine particles, which can be carried out by airborne particulate matter including soil dust, pollen, ashes, etc [3], deposition of soluble materials on dust particle surface can take place by coagulation between soil dust and fine particles; or by occurrence of heterogeneous reactions between soil particles and reactive gases, as example CaCO$_3$ particles can react with HNO$_3$ or HCl to form Ca(NO$_3$)$_2$ or CaCl$_2$, these products are highly soluble materials [4].

Examples of dust particles analysis [3, 4, 5] agree in establish that dust particles can be a place to form new products, as well as being formed by materials which can be either water soluble or acid soluble. Also quantification was done for cations like Zn, Pb, Mn, Fe, Cr, Mg, V, Cu, Ti, Ca, Al, Na, K, NH$_4$, as well as anions like F, Cl, Br, NO$_3$, SO$_4$.

Experimental details clearly establish that dust components can be extracted by water or by an acid solution, Tomohiro Kyotani [5] also provides a flow diagram in which it is considered an additional extraction for organic components. Also this author point out that a large part of ions like aluminum can not be water extracted; as well as can occur that highly water soluble ions like Ca an Mg, could render higher amounts when they are extracted by an acid solution. Also, there is evidence of salt clusters is provided as SEM-micrographs [4].

EXPERIMENTAL DETAILS

Dust samples were collected from roofs of 3 buildings at downtown, sites were chosen based on different environment exposition and taking care that distance between them does not exceed 500 meters. The first corresponds to the Engineering Board Association (EBA), which is a two level building located at a highly transited avenue, on which run urban buses as well as commercial and private vehicles, also, it is surrounded by buildings whose height is one or two levels, this fact provides a wide open space around the EBA building; the second one corresponds to the Colonial Hotel (CH), which is a 4 level building with transit restricted to compact and medium vehicles circulating only by the south side and it is surrounded by buildings with diverse heights, also it is an area with a high population density, around it most of the places provide food service on site; the third one corresponds to the Saint Agustin temple (SAT), which is located in an avenue with similar restrictions to the CH, but with transit circulating on the east and north sides, also it is surrounded by tall buildings mainly at the north and northeast, in the nearby area there are many commercial places and business offices; also there are some places providing food service but fewer than the ones around the CH.

Analysis of collected dust samples was done in three ways: direct sample analysis, water soluble components, and acid soluble components.

Direct sample analysis

In this step, dust particles were analyzed for: 1) oil and grease content determined by the soxhlet technique; 2) elemental analysis detecting C, N, H with a Perkin Elmer 2400

CHN Elemental Analyzer; 3) mineral carbonate fraction determined by the calcimetry technique.

Water soluble analysis

For this stage, the procedure was accomplished by a common technique in soil analysis, in which an analytical sample is prepared by mixing dust with distilled water (1:2); after 24 hours contact, these samples were filtrated and used to measure pH and conductivity with Conductronic Instruments (10 pHmeter, 8C conductimeter); the liquid was analyzed by gravimetric methods for chloride, alkalinity, and hardness content; determination of sulfate, nitrate, and reactive sulfide was done with a 2500 Hach visible spectrophotometer; while soluble cations were determined in a GBC 932 atomic absorption spectrometer

Acid soluble analysis

For non-soluble components an acid digestion was performed by placing 1 g of sample for acid digestion with 100 ml of 5% nitric acid and 2% HCl solution, this procedure allowed metals solubilization, and they were detected with a GBC 932 atomic absorption spectrophotometer.

DISCUSSION

In order to estimate if vehicle transit and human activities influence dust properties, it was considered to set a 30 minutes period for counting the number of vehicles circulating by the chosen places. Results are shown in figure 1. As can be observed the number of small vehicles is similar for the three places, while the number of medium size vehicles is similar only for CH and SAT, and it is slightly higher for the EBA; which is the site accounting for large vehicles. In some way vehicle emissions should be contributing for dust to acquire organic content which can be determined by parameters such as carbon, nitrogen, hydrogen. An indirect way of assessing the organic content, it is through testing dust samples for oil and grease, and its presence can be confirmed by elemental analysis.

Direct sample analysis

Oil and grease content determined by the soxhlet technique clearly reflects the activities taking place at each site, results are shown in figure 2, as can be observed the CH with higher population in its surroundings has the highest content (249 ppm) of oil and grease; otherwise, the EBA has lower content (21 ppm) which is less than 10% of the CH, even though EBA is located at an avenue with high traffic density, and the lower concentration corresponds to SAT (5 ppm). Detected values in the EBA and SAT clearly can be attributed to vehicle emissions, while the high amount detected at the CH obviously must have an alternative origin, because vehicle transit at CH is similar to SAT.

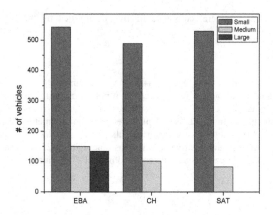

Figure 1. Estimated vehicle density at Engineering Board Association (EBA), Colonial Hotel (CH) and San Agustin Temple (SAT), during a 30 minutes period.

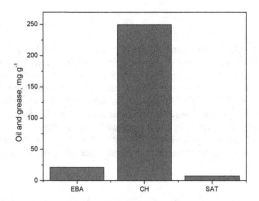

Figure 2. Oil and grease content in dust samples collected at Engineering Board Association (EBA), Colonial Hotel (CH) and San Agustin Temple (SAT).

Based on oil and grease results, elemental analysis was run for the EBA and SAT sites. Analytical results   are shown in figure 3. It was found that C, N, H concentrations at the EBA are greater than those found in SAT dust. These results agreed with the expectations from oil and grease analysis. In general, C, N and H contents should be higher at the place having greater movement of vehicles; and, of course the fact of EBA having an important number of large vehicles circulating by this place should exert a strong impact on the organic content because most of urban buses use either diesel or gas.

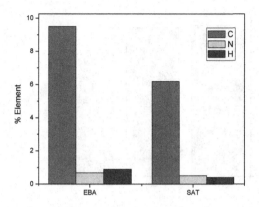

Figure 3. Elemental analysis results for carbon (C), nitrogen (N) and Hydrogen (H) in dust samples from Engineering Board Association (EBA) and San Agustin Temple (SAT) places.

As it was assumed from 2009 rainfall statistical correlations [2] ions present in rainwater can have a common origin if they correlate positively as example marine components like sodium and chloride correlate with a 0.6 factor, otherwise sodium and carbonate correlate with -0.02 factor; from the same report calcium, magnesium, carbonate and sulfate should have a common origin which is different from rainwater since the correlation between CO_3 and Ca is 0.83, while CO_3 and Mg is 0.98; therefore, atmospheric dust should have alkaline properties [1]. For collected samples carbonate content was determined by calcimetry, Results are shown in Figure 4, as it can be observed carbonate content is similar for EBA and CH, and a little lower for SAT, although it can be affirmed that carbonate content is higher than 30% weight for all places.

Soluble components

With the analytical sample filtrate pH and conductivity were determined, results are shown in figure 5, from conductivity results it is obvious that ion content is high for all samples, and even though there is not a clear pattern in conductivity, since it can be seen that higher conductivity corresponds to CH, the taller building, followed by the EBA and the lower one is SAT, if this parameter were expressed in percentage considering CH as 100%, then EBA would have an 85 % and SAT 25% of ionic content. Although this response does not correspond to the pH trend since the EBA with lower conductivity exhibit a pH of almost 8, while CH with higher conductivity has a pH of 6; and even though SAT has lower conductivity its pH is around 6.

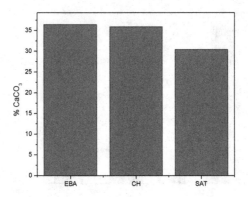

Figure 4. Carbonate content determined by calcimetry in dust samples from Engineering Board Association (EBA), Colonial Hotel (CH) and San Agustin Temple (SAT) places.

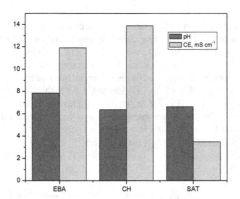

Figure 5. pH and conductivity results determined in analytical samples prepared with dust from Engineering Board Association (EBA), Colonial Hotel (CH) and San Agustin Temple (SAT) places.

Figure 6 shows the results for HCO_3^- determined as methyl orange alkalinity (MO alc) and CO_3^{-2} determined as total hardness, as can be observed MO alc is similar for EBA and CH, and it is almost twice in SAT; otherwise total hardness is higher for the EBA, and a little lower for CH and SAT, these have similar values. Comparison of these CO_3^{-2} values, which are the soluble fraction of the carbonate mineral content (Figure 4) it can be affirmed that almost 10% of the mineral content is easily solubilized.

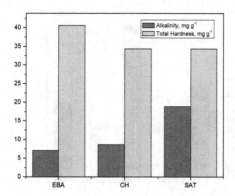

Figure 6. Water soluble concentrations for HCO_3^- determined as alkalinity (M-O Alc), and CO_3^{-2} determined as Total hardness for Engineering Board Association (EBA), Colonial Hotel (CH) and San Agustin Temple (SAT) places.

Considering that sulfur presence is mainly in the sulfur dioxide form, which can be produced by anthropogenic activities involving fuel combustion; but also, there are non anthropogenic sources like volcano emissions and biological decay of organic matter. SO_2 like many other gaseous pollutants reacts to form particulate matter step that can occur by addition of an HO* radical producing the $HOSO_2$* specie which eventually will be converted to a SO_4 form. Also, it is known that SO_2 in presence of limestone or dolomite will produce gypsum ($CaSO_4$) [6]. Because dust particles have an important amount of carbonate concentration, then sulfur can be present as either reactive sulfur (S react) or sulfate (SO_4).

Figure 7 shows the results for sulfate (SO4) and reactive sulfur (S react), also it have been included chloride (Cl) because it is a powerful oxidizing agent which, once released in the atmosphere, can be dissolved in atmospheric water droplets yielding either hydrochloric or hypochlorous acid; so far chloride concentrations seem to be related with occurrence of reactive sulfur transforming into sulfate. From these results it is clear the CH place has low sulfate presence associated to high chloride concentration, it seems that in some way the simultaneous occurrence of chloride in presence of high organic content (Figure 2) inhibits the conversion of sulfide to sulfate; otherwise in EBA and SAT, both with low organic content, occurs that chloride concentrations are lower than the one at CH, in both sites sulfate formation seems to be proportional to carbon percentage. Although, sulfate conversion seems not being proportional to the amount of reactive sulfur, because at the EBA sulfate is almost twice the amount observed at SAT, but reactive sulfur at EBA is almost 60% of the one present at SAT.

Main cations concentrations are reported in figure 8, these are calcium, and iron. Reported concentrations show how mineral content can be high, but soluble fraction is low since soluble calcium less than 30% , as well for iron soluble fraction is less than 1%.

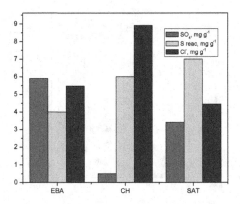

Figure 7. Soluble concentrations of sulfate (SO4), reactive sulfur (S react) and chloride (Cl) for Engineering Board Association (EBA), Colonial Hotel (CH) and San Agustin Temple (SAT) places.

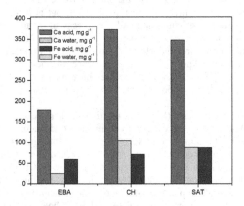

Figure 8. Comparison of water soluble and acid digestion concentrations for calcio (Ca), and iron (Fe) in dust samples from Engineering Board Association (EBA), Colonial Hotel (CH) and San Agustin Temple (SAT) places.

Figure 9 shows the concentrations for aluminum and copper, these concentrations are lower than calcium and iron, but as can be observed aluminum is not water soluble, and opposite to this response is the one exhibited by copper since above 60% of mineral copper is easily solubilized.

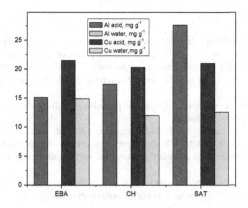

Figure 9. Comparison of water soluble and acid digestion concentrations for aluminum (Al), and copper (Cu) in dust samples from Engineering Board Association (EBA), Colonial Hotel (CH) and San Agustin Temple (SAT) places

Figure 10 shows the concentrations for manganese, lead and zinc, as it can be observed that soluble manganese is about 15% of its mineral concentration, also zinc is present in the mineral but is not water soluble; Opposite behavior is exhibited by mineral lead, which even though it is in low concentration, its solubility reaches almost 70% of the mineral content.

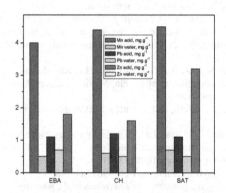

Figure 10. Comparison of water soluble and acid digestion concentrations for manganese (Mn), lead (Pb) and zinc (Zn) in dust samples from Engineering Board Association (EBA), Colonial Hotel (CH) and San Agustin Temple (SAT) places

CONCLUSIONS

This study provides information on dust characteristics making evident that anthropogenic activities are a source for organic matter presence in dust particles. A comparison of sites with similar vehicle density can be used as criteria to discriminate the origin of organic carbon; also, higher vehicle density is reflected in higher content of C, H, N.

Mineral carbonate content is around 30% for all particles, and from this only 10% is water soluble determined as total hardness.

Adsorbed materials can undergo chemical reaction, and it was observed that conversion of reactive sulfide to sulfate is highly dependent on the organic content, since higher carbon content correlates with a higher presence of sulfide and lower amounts of sulfate.

Also a comparison of acid digested metals with water soluble ones allows detection of non-soluble ones like Al, Fe, and Zn; as well as highly soluble ones like Ca, Pb.

ACKNOWLEDGMENTS

The authors would like to thank to Priest Miguel Chávez Garcilazo responsible of San Agustin Temple (SAT), Engineering Board Association (EBA), and Colonial Hotel (CH) for facilities to collect samples. Also, authors are thankful for the financial support provided by SEP through the PROMEP program IDCA 6941.

REFERENCES

1. M. Teutli, E. León, J. Cerna, A. C. Ruíz, Basalt chemical weathering at Puebla historical buildings, in 2nd *Latin-American Symposium on Physical and Chemical Methods in Archaeology, Art and Cultural Heritage Conservation. Selected papers*, J.L. Ruvalcaba, J. Reyes, J. Arenas, A. Velazquez (eds.), Universidad Nacional Autonoma de Mexico, Universidad Autonoma de Campeche Instituto Nacional de Antropologia e Historia, Mexico, 2010, 172-175.
2. M.Teutli León, G. Jiménez Suárez, A.A. Peláez Cid, J. Lozano Mercado, A. E. Posada Sánchez. Rainfall chemical composition at Puebla, México. *Enlace Químico*, vol 2 (9), 2010. http://www.revistaequim.com/
3. C. Buourotte, M. C. Forti, A. J. Melfi, and Y. Lucas. *Water, Air and Soil Pollution*, 170, (2006) 301-316.
4. Y. Tobo, D. Zhang, A. Matsuki, Y. Iwasaka. Asian dust particles converted into aqueous droplets under remote marine atmospheric conditions, *PNAS* 07, 42 (2010) 17905-17910. http://www.pnas.org/cgi/doi/pnas.1008235107
5. T. Kyotani, M. Iwatsuki. *Analytical Sciences*, 14 (1998) 741-748.
6. S.E. Manahan. Gaseous inorganic pollutants in *Environmental Chemistry*, S.E. Manahan ed., Lewis Publishers, Boca Raton, Florida, 2000, 329-350.

Mater. Res. Soc. Symp. Proc. Vol. 1374 © 2012 Materials Research Society
DOI: 10.1557/opl.2012.1391

Additions of Minerals in Clays of Morelia Region, Mexico: Effects on Volumetric Stabilization and Color

W. Martínez Molina[1], J. L. Ruvalcaba Sil[2], E. M. Alonso Guzmán[1], A. Flores Rentería[1], M. Manrique Ortega[2], A. A. Torres Acosta[3]

[1]Facultad de Ingeniería Civil de la Universidad Michoacana de San Nicolás de Hidalgo, Edificio F de Ciudad Universitaria, Avenida Francisco J. Mújica s/n, Colonia Felícitas del Río, Morelia, Michoacán, Mexico CP 58040, Mexico.
  e-mail: wilfridomartinezmolina@gmail.com; eliamercedesalonso@gmail.com
[2]Instituto de Física, Universidad Nacional Autónoma de México UNAM, Mexico.
  e-mail: sil@fisica.unam.mx
[3]Universidad Marista de Querétaro, Querétaro Centro Histórico, CP 76000, Querétaro, Mexico.

ABSTRACT

Clays were used intensively in cultural heritage's monuments and objects. Conservation procedures can be performed specifically for earthen materials using stabilized clays, considering that the aesthetic features must be preserved in order to avoid drastic differences and the lost of their patrimonial value.

This work presents the study of the mechanical behavior of clay stabilized with different materials following the norm ASTM D 6276 – 99ª, for lime stabilization. The effects of other stabilizers on the clay were studied as well. For these purposes, lime, gypsum, Portland cement (type II), sodium hydroxide, and dehydrated cactus fibers of white *cactus opuntia* in concentrations of 2, 4, 6, 8 and 10 wt% were added to a clay from Morelia region.

Atterberg limits were determined to calculate the linear and volumetric stabilization. The best volumetric stabilization values were chosen to prepare samples to measure the mechanical behavior under compression, tension and flexion strengths. Colorimetric measurements were also performed on the stabilized clays to determine the best preparation with the most suitable aesthetic qualities to perform conservation treatments on monuments and cultural heritage constructions made with earthen materials.

The highest values for compression were observed for gypsum and mucilage additions while the highest tension was obtained for mucilage ones. Gypsum addition had the bigger rupture module under flexion. On the other hand, the color of the stabilized clay is closer to the original clay color for cement, lime and mucilage preparations.

INTRODUCTION

In Michoacan region, Mexico, the Purepecha people developed, especially at the shores of the lakes of Patzcuaro, Zirahuen and Cuitzeo, where a significant concentration of historical constructions and monuments made of earthen materials (adobe) remain nowadays. Adobes were prepared using clays, very abundant in the

Pacific Coast of the country. The clays studied in this work are from the Santiago Undameo quarry, Michoacan, Mexico, at the shore of the Cointzio dam.

Many of the historical buildings and monuments are still used for their original function: Churches, hospitals and housing. They are exposed to environmental conditions and the corresponding deterioration. For this reason, they required conservation strategies and procedures, restoration materials and maintenance for a medium wet weather.

Clays presents a laminate structure and layers and their structures present interlayer spaces for water absorption that may give rise to volumetric expansions and contractions [1, 2].

The Unified Soil Classification System, USCS classifies clays as high and low compressibility, HC and LC, and the expansion index is also another criterion for the mineralogical classification [3, 4]. Smectites belong to young clays with high expansion/contraction index [5]. A general classification includes montmorillonites, illites, kaolinite, vermiculites and chlorites. A formal definition for clays establishes that it is a material composed by microscopic flakes particles of mica and other minerals of dimensions smaller than 0.002 mm. Clays are composed by hydrated alumina-silicates with Mg, Fe and other metals [6, 7].

Soils composed by clays are hygroscopic and the water absorption or desorption produces significant volumetric changes [8, 9]. To reduce these effects other materials may be introduced in the layer structure to avoid water absorption. Different kind of minerals, organic and inorganic materials may be used for this purpose. Macroscopic and microscopic vegetable fibers may also be used [10].

This research focuses on materials that work at crystallographic level to modify the clay properties [11], the studied additions include construction grade lime [12], gypsum, Portland cement type II, sodium hydroxide and dehydrated cactus fibers of white cactus opuntia. The modified clay properties were compared to the control clay sample [13, 14] to verify the mechanical behavior.

EXPERIMENTAL

Clays were extracted from the sources of quarry clay of Santiago Undameo, Michoacan, Mexico, at the shore of the Cointzio dam. This material was separated by a 40 ASTM mesh, materials retained in 40 mesh are sand, and material minor to 40 mesh are a kind of soil [15]. Water is added to the separated soil to reach the liquid limit (LL), the plastic limit (PL), plastic index (PI), contraction limit (CL) and linear contraction (LC). After the quantification of these parameters, it is possible to determine the type of soil and the volumetric stability. The parameters are known as Atterberg limits for soil classification of USCS.

First the clay was characterized using X-ray Diffraction (XRD). The powdered clay was separated by the 400 mesh. Then analyzed with a Bruker diffractometer AXS model D8 Advance with a Linx detector and a X-ray tube of Cu.

Testing for soils classification was carried out by triplicate for the control clay sample as well as for each of the stabilized clays as a function of the concentration of the added material. The clays were stabilized using usual and traditional materials such as construction grade lime, gypsum [16], Portland cement [17] type II, sodium hydroxide and materials described in the historical sources like de-hydrated cactus fibers of white cactus *opuntia indica*. These fibers when mixing with water produce calcium gels that after solidification give rise to calcium minerals such as oxalates, whewellite, and weddellite.

Using the stabilized clay data [18], some samples were prepared to measure their mechanical properties. There are not universal procedures for this testing. In soil mechanics engineering it is enough to find the addition wt% that produce the lower volumetric changes independently of the mechanical resistance.

The norms used to prepare and test the samples were the ASTM for lime mortars. For compression testing, specimens of 5 cm of edge were prepared, while for the flexion tests a charge was applied at the middle third with dimension of 4cm x 4 cm x 20 cm. For the tension test, briquet specimens were prepared of 10 cm length x 1.8 cm height and 3.5 cm width. The specimens were prepared in the laboratory adding mineral oil to avoid the mold's adhesion.

Different aging were also tested monitoring the behavior after the setting of the material. The aging times were of 3, 7, 14, 21 and 28 days. Nevertheless, the 3 days aging samples were not used because they remained attached to the mold. About 50% of the specimens fractured when drying and were discarded.

The cubic and prisms specimens were tested in a Tinius Olsen universal machine with maximum load of 50 tons and increments of 0.1 kg. The cubes were set on a layer of silica sand previously separated by a 16 mesh and retained by a 30 mesh. The charge was applied until fracturing. The briquets were set at the clamps in a mechanical machine Michaellis until the fail at their geometric medium centers.

Colorimetric measurements were done in both faces of the specimens in three regions with an Ocean Optics USB4000 spectrometer with a probe and optical fibers of 400 μm diameter for the 300 to 1000 nm range, averaging 20 spectra of 50 ms. A halogen source from Ocean Optics was also used. The measurements were carried out perpendicular to the surface. The spectrometer was calibrated by a white color reference (WS-1 diffuse reflectance standard) with the Spectrasuite software from Ocean Optics. The measurements were averaged and the standard deviation was calculated for each clay preparation.

RESULTS

The XRD measurements indicate that the clay is a Kaolinite with high temperature crystals of quartz (Figure 1).

From the stabilized clays testing, it was observed that none of the cactus mucilage samples reached the required pH of the norm ASTM D 6276, with lower values. This happened as well for the gypsum. Only the additions of sodium hydroxide, Portland cement and lime at 6 wt% and higher concentration accomplished the norm with a pH around 12.4.

Atterberg limits were measured: the mixture of clay and lime, and clay and cement, and clay witness presented a significant improvement with a good control on the volumetric changes. In contrast, for the mixture of clay and calcium hydroxide from 6 wt%, a layer of crystals formed at the surface of the specimens with a consequent fracturing and breaking of the sample.

For the mechanical testing of compression, tension and flexion, only the 6 wt% mixtures were used for lime, cement, gypsum since the change in the color was significant by comparison to the original clay. For the other samples of calcium hydroxide mixture and the cactus fibers [10] the 2 wt% concentration was used because some crystallization appeared al higher concentrations. Also most of the calcium hydroxide mixture samples were fractured and broken during the drying process.

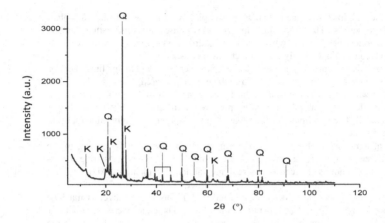

Figure 1. XRD pattern for the clay from Santiago Undameo quarry, Michoacan, Mexico. Kaolinite (K) and quartz (Q) reflections are shown.

pH Measurements

The pH values were monitored to determine the mixtures stability that occurs when the liquid with the volumetric stabilizer reach a pH of 12.4. (Standard ASTM D 6276). Table 1 shows the corresponding measurements. The stabilizer and the concentrations in agreement with the norm are indicated. This happens only for lime (4-6 wt%), sodium hydroxide (2-6 wt%) and Portland cement Type II (8 wt%).

Table 1. pH measurements for the used stabilizers as a function of the lime wt%. The preparations in closer agreement with the norm ASTM D 6276 are indicated.

| Stabilizer wt% | pH | | | | |
	Cactus fibers	Sodium Hydroxide	Gypsum	Lime	Portland cement
2	5.8	12.3	5.6	11.3	10.8
4	5.9	12.5	5.8	12	11.5
6	6.2	12.6	5.9	12.3	11.7
8	6.5	12.9	6.2	12.6	12.1
10	7.8	13.7	6.6	13.2	12.9

Atterberg limits

The measured Atterberg limits: liquid limit (LL), the plastic limit (PL), plastic index (PI), contraction limit (CL) and linear contraction (LC) as well as the USCS for the mixtures are shown in tables 2 to 6.

218

Mechanical testing

The compression resistance measurements of the cubes are shown in Figure 2 while the resistances to flexion and to the tension data are shown in Figure 3 and 4, respectively.

Table 2. Atterberg limits for the mixture of clay and lime.

Limits	Natural clay	Lime stabilizer				
		2 wt%	4 wt%	6 wt%	8 wt%	10 w%
LL	66	51.2	44.8	43.9	43.64	43.2827
PL	24.8	25.8	35.38	39.47	30.18	36.46
PI	41.2	25.4	9.42	4.43	13.46	6.82
CL	14.12	17.43	27.21	27.67	28.56	40.46
LC	17.5	13.1	7.9	6.2	6.01	4.2
SUCS	CH	CH	ML	ML	CL	ML
Δ C water%	51.88	33.77	17.59	16.23	15.08	2.8227

Table 3. Atterberg limits for the mixture of clay and gypsum.

Limits	Natural clay	Gypsum stabilizer				
		2%	4%	6%	8%	10%
LL	66	60.8	57.5	57.5	57.127	49.832
PL	24.8	22.32	17.43	25.44	16.08	19.91
PI	41.2	38.48	40.07	32.06	41.04	29.92
CL	14.12	14.95	15.45	19.53	27.45	16.06
LC	17.5	15.9	15.5	14.55	15.6	20.27
SUCS	CH	CH	CH	CH	CH	CL
Δ C water%	51.88	45.85	42.05	37.97	29.677	33.772

Table 4. Atterberg limits for the mixture of clay and cactus fibers (mucilage).

Limits	Natural clay	Cactus mucilage stabilizer				
		2%	4%	6%	8%	10%
LL	66	68	68.8	73.5	82	91.345
PL	24.8	22.31	19.94	15.22	19.26	19.66
PI	41.2	45.69	48.86	58.28	62.89	71.68
CL	14.12	20.61	7.43	14.28	28.82	16.82
LC	17.5	16.1	16.4	16.2	17.9	19.3
SUCS	CH	CH	CH	CH	CH	CH
Δ C water%	51.88	47.39	61.37	59.22	53.18	74.525

Table 5. Atterberg limits for the mixture of clay and Portland cement Type II.

Limits	Natural Clay	Portland cement stabilizer				
		2%	4%	6%	8%	10%
LL	66	50.5	49	47.8	47.17	50.16
PL	24.8	29.69	22.82	30.5	30.9	39.57
PI	41.2	20.81	26.18	17.3	10.27	10.59
CL	14.12	15.34	20.67	28.04	35.32	35.77
LC	17.5	13	12.3	7	5.6	4.9
SUCS	CH	MH	CL	CL	ML	ML
Δ C water%	51.88	35.16	28.33	19.76	11.85	14.39

Table 6. Atterberg limits for the mixture of clay and sodium hydroxide.

Limits	Natural Clay	Sodium hydroxide stabilizer				
		2%	4%	6%	8%	10%
LL	66	59.2	59	62.9	53.32	53.86
PL	24.8	53.91	30.45	30.27	24.91	28.4
PI	41.2	5.29	28.55	32.63	28.41	25.46
CL	14.12	15.91	19.37	28.94	21.87	33.69
LC	17.5	16.1	13.48	11.1	9	6.8
SUCS	CH	MH	CH	CH	CH	CH
Δ C water%	51.88	43.29	39.63	33.96	31.4519	20.17

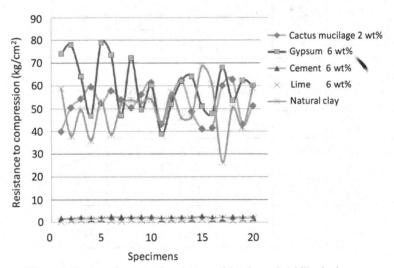

Figure 2. Compression strength resistance of the clay and stabilized mixtures.

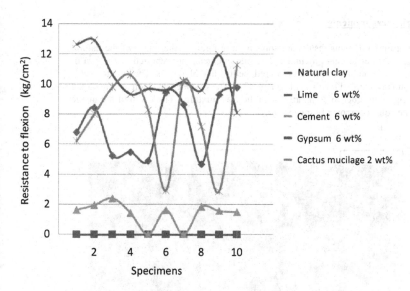

Figure 3. Flexion strength resistance of the clay and stabilized mixtures

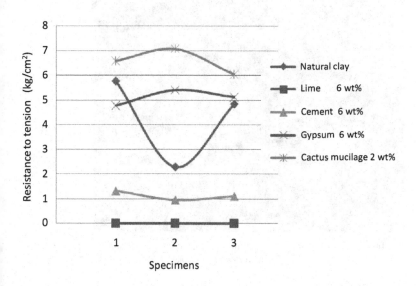

Figure 4. Tensile strength resistance of the clay and stabilized mixtures.

Colorimetric measurements

The color measurements results are condensed in figure 5, the image of the natural clay and the mixtures are also included. The measurements are reported in the $L^*a^*b^*$ CIE system, corresponding in the figure to red, yellow and blue lines. The average color of the natural clay is $a^* = 11.1$, $b^* = 11.8$, $L^* = 43.2$. These values are indicated as doted lines in the figure in order to compare it with the clay mixtures. The width of the image represents the color component range variation.

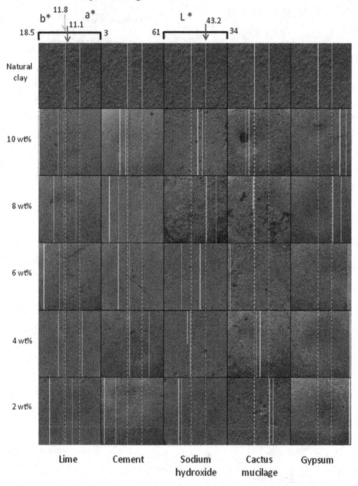

Figure 5. Color measurement schema for the clay mixtures. The red, yellow and blue lines correspond to a*, b* and L*, components. The color data are indicated for the natural clay.

The clay preparations are practically all different by comparison to the natural clay. Nevertheless, the colors of the mixtures are similar to the natural clay color for the 4 wt% of lime, cement, sodium hydroxide and cactus mucilage, but 10 wt% of cement and sodium hydroxide, and 8 wt% of cactus mucilage, give rise also to a comparable color.

DISCUSSION

The experimental data show that is possible to follow the norm ASTM D 6276 and to produce the reaction at a pH of 12.4 and 25 °C, in order to stabilize the clay preparation, nevertheless the mechanical resistance did not increase.

Only the samples of sodium hydroxide at 2 wt%, Portland cement and lime at 8 wt% agree the norm ASTM D 6276, gypsum and cactus mucilage preparations did not fit the norm.

The additions of Portland cement and lime at 6 wt% were the best to control the volumetric changes but these samples presented the lowest resistance to tension, compression and flexion, even lower values than the natural clay.

In contrast the best mechanical properties were obtained for the 6 wt% gypsum sample with and increment for the resistance to tension, compression and flexion by comparison to the natural clay.

On the other hand, for the cactus mucilage samples at 2 wt%, there is a moderate increment in the resistance compression and flexion. This sample has the highest values of resistance to tension.

The limits of consistence were measured and the best preparation corresponds to the sample of lime at 10 wt% with a good control on the volumetric changes. The color of this sample is more yellow and lighter. Portland cement samples presented a similar behavior.

The additions of sodium hydroxide and cactus mucilage produced at 6 wt% and higher concentrations, a white layer. Then these preparations are not suitable for consolidation purposes.

In general, when the concentration of the additions were higher than 6% there is an important change in the color of the samples by comparison to the natural clay. Gypsum samples keep a similar color despite its concentration increases.

The additions of cement, sodium hydroxide and cactus mucilage are the most similar in color to the natural clay.

CONCLUSIONS

The studied additions to the natural clay modify its properties and color. There is an improvement in the mechanical behavior as well as in the volumetric changes. Considering these aspects, the most suitable preparations of clays correspond to 4 to 6 wt % of Portland cement and 2 to 4 wt% of cactus fibers of white *opuntia*.

These materials are suitable for further cultural heritage conservation testing and consolidation treatments. These kinds of studies are necessary to determine more suitable materials for conservation and with a longer duration.

ACKNOWLEDGMENTS

This research has been supported by the CIC of the Universidad Michoacana de San Nicolás de Hidalgo and the grants from CONACYT MOVIL II 131944 and DGAP-PAPIIT UNAM IN403210 ANDREAH project. Authors thank M. Aguilar Franco for the XRD analysis.

REFERENCES

1. D. Braja M., *Fundamentos de ingeniería geotécnica*, Thomson Learning, Mexico, 2006.
2. J. M. Domínguez, I. Schiffer, *Las arcillas el barro noble*, Ciencia 169, Fondo de Cultura Económica. Mexico, 1999, 25-39.
3. M. El-Sadek Abdel Rahman Ouf, *Stabilisation of clay subgrade soils using ground granulated blastfurnace slag*, Thesis of Doctor of Philosophy, School of Civil Engineering University of Leeds, U.K., 2001, 65-80.
4. E. García Romero, M Suárez Barrios, *Las arcillas: propiedades y usos*, Universidad Complutense y Universidad de Salamanca, Madrid, 2002.
5. Y. Gurtug, *Prediction of the compressibility behavior of highly plastic clays under high stresses*, Applied Clay Science 51 (2011) 295-299.
6. S. Horpibulsuk, R. Runglawan, A. Chinkulkijniwat, Y. Raksachon, A. Suddeepong, *Analysis of strength development in cement-stabilized silty clay from microstructural considerations*, Construction and Building Materials 24 (2010) 2012-2020.
7. K.M.A. Hossain, M. Lachemi, S. Easa, *Resources, Conservation and Recycling* 51 (2007) 715-725.
8. E. Juárez Badillo, A. Rico Rodríguez, *Mecánica de Suelos*, Tomo I, Fundamentos de la Mecánica de Suelos, Editorial Noriega Limusa, Mexico, 2006.
9. *Manual de prácticas de mecánica de suelos*, Séptimo Semestre, Laboratorio de Materiales "Ing. Luís Silva Rúelas", Facultad de Ingeniería Civil, Universidad Michoacana de San Nicolás de Hidalgo, Morelia, Michoacan, Mexico, 2010.
10. C. Márquez Alonso, *Plantas Medicinales de México II, Composición, Usos y Actividad Biológica*, Universidad Nacional Autónoma de México, Mexico, 1999.
11. W.H. Matthews, *Geology made simple*, Services Company Ed., 1977, 63 - 65.
12. M.A. Olguín Domínguez *Efectos Mecánicos de la estabilización volumétrica de montmorillonita con CaSO₄*, Tesis de Licenciatura, Facultad de Ingeniería Civil, Universidad Michoacana de San Nicolás de Hidalgo, Morelia, Michoacan, Mexico, 2008.
13. S.M. Rao, M, P. Shivananda, *Geotechnical and Geological Engineering* 23 (2005) 309-319.
14. A. Seco, F. Ramírez, L. Miqueleiz, B. Garcia, *Applied Clay Science* 51 (2011) 348-352.
15. F. V. Villaseñor, *Estudio de Mecánica de Suelos y Cálculo Estructural de los Cinemas Gemelos Plaza Hidalgo en León Gto.*, Dissertation, Facultad de Ingeniería Civil, Universidad Michoacana de San Nicolás de Hidalgo, Morelia Michoacan, Mexico.
16. I. Yilmaz, B. Civelekoglu, *Applied Clay Science* 44 (2009) 166-172.
17. R.N. Yong, V. R. Ouhadi, *Applied Clay Science* 35 (2007) 238-249.
18. J. Wesley Parker, *Evaluation of laboratory durability tests for stabilized subgrade soils*, Thesis of Master of Science, Department of Civil and Environmental Engineering, Brigham Young University, Provo, 2008, 7-26.

Biomaterials Topics

Mater. Res. Soc. Symp. Proc. Vol. 1374 © 2012 Materials Research Society
DOI: 10.1557/opl.2012.1392

Trace Element Analysis of Bone from Past Populations in the Peninsula of Yucatan

Saul Chay[1], Mónica Rodríguez[1], Patricia Quintana[2], Vera Tiesler[1]
[1]Facultad de Ciencias Antropológicas, Universidad Autónoma de Yucatán, Mérida Mexico. e-mail: sachayvela@hotmail.com
[2]Departamento de Física Aplicada, Centro de Investigación y de Estudios Avanzados (CINVESTAV) Mérida, Yucatán, Mexico.
e-mail: pquin@mda.cinvestav.mx

ABSTRACT

This dietary study compares concentrations of trace elements in human skeletal series from the municipal cemetery of Xoclán, in Mérida, Yucatan, and a skeletal collection that was donated by the Yucatecan State Justice Department (PGH). The results from these modern samples are to be compared to those obtained from human collections from a colonial cemetery from Campeche and the pre-Hispanic Maya site of Xcambó. Our results indicate that the archaeological series show higher concentrations of Sr compared to the modern populations, both of which showed very similar values. Zn concentrations were similar when the modern values were compared to those derived from the colonial series from Campeche. Xcambó's population, in turn, shows a high degree of variability in Zn values, which may be due to diagenetic contamination.

INTRODUCTION

Trace element analyses were the first quantitative approach to the study of ancient diet through osteological evidence from archaeological contexts. This technique measures the mineralized substrate of bone (hydroxyapatite). That approach is useful therefore to identify dietary macrocomponents in past societies. First applied to the study of radioactive environmental contamination [1], the method is founded on the natural concentration of elements like barium (Ba) and strontium (Sr) in geological strata [2]. These elements are passed on to plants and from there to those animals that consume plants. Given that both elements possess the same chemical position as calcium (Ca), they are capable of substituting it inside the hydroxyapatite molecule. The principle of "biopurification" which prescribes that Sr and Ba decrease their concentrations in bone tissue as the organism rises in the food chain [1], imply that their concentrations are lower in organisms that feed on animals than vegetarian individuals. Complementary measures taken of additional elements like iron (Fe), zinc (Zn) and potassium (K) provide complementary dietary scores, although results are invalidated, especially since all three of those elements are controlled directly by metabolic functions and only indirectly reflect dietary intake [3].

This work compares elemental concentrations of Sr and Zn between modern and past osteological human samples from the Peninsula of Yucatan, each collection possessing its specific environmental, historic and cultural context, which we intend to interpret in the following in terms of diet, living conditions and diagenetic factors.

SAMPLING

We selected bone samples from three different contexts, all of which are located on the Yucatecan peninsula. Firstly, the documented modern cohort of Merida comes from the Municipal Cemetery of Xoclán. These individuals died and were buried between 2000 and 2005. A second modern skeletal series has been donated by the State Depatment of Justice (PGJ), and includes individuals from all over the state of Yucatan. The third, historical human series under study derives from a colonial cemetery in downtown Campeche. This last cohort includes blacks, *mestizos*, natives and Europeans and dates to the XVI and XVII century [4]. In order to have a more complete comparative prospect, we compare our results with elemental data from the pre-Hispanic site of Xcambó, reported by Tejeda and colleagues [5]. The occupation of the site occurred between 350 and 700 A.D. This Classic period coastal settlement is located on an artificial island in the northeastern fringes of the State of Yucatan [6]. Here, the ancient community produced salt and was active in exchange [7].

Figure 1. Regional map showing sites for this work.

MATERIALS AND METHODS

Samples were taken from sections of compact bone in the femur. Each of them was freed of its outer layers with a Dremel saw to eliminate possible diagenetic contamination. Each sample was then dried at 80 and 100C, crushed and compressed into pellets, each weighting 0.5 g. Each pellet was measured five times per sample.

The analysis was conducted in the Department of Applied Physics in the CINVESTAV Mérida according with the X-ray fluorescence (XRF) technique, focusing on the strontium (Sr) and zinc (Zn) as dietary indicators. For this purpose, we employed a spectrometer (Jordan Valley EX-6600) equipped with a Si (Li) detector holding 20 mm^2 of radiation area, a resolution of 140 eV at 5.9 keV, and operating at a maximum of 54 kV and 4800 μA. The trace elements were scrutinized with an exchangeable detection spot within a environmental pressure of one atmosphere [8].

To obtain calibration curves, we used hydroxyapatite standard samples of SRM1400 y SRM1486. Calibration curves were generated also to measure Sr and Zn as trace elements. Measures obtained from the highly concentrated elements calcium (Ca) and phosphorus (P) were also obtained and expressed as percentages. The measured values were averaged, normalized and converted to a logarithmic scale by using the following formula:

$$X/Ca = (X_{ppm}/Ca_{ppm})$$

This formula describes the normalized value of an element, expressed as parts per million, and divided by the total concentration of calcium, as expressed in ppm. This value is imported into a logarithmic scale to facilitate comparisons of extremely different value ranges.

RESULTS

Correlations between Sr/Ca and Zn/Ca concentrations evidence huge differences between the archaeological and forensic populations scrutinized here (Table 1). Archaeological samples present higher levels of Sr in comparison with forensic individuals, with the exception of a single individual from PGJ collection, found in a marine context (Figure 3). Zn is more stable and it does not show significant differences between populations, although it appears to exhibit a greater variability among individuals from Xcambó [5]. There is no strong correlation between the concentrations of both elements, with the only exception of Xcambó´s sample, which we think is due, more than to ancient dietary intake, to the diagenesis that prevails in this set.

Table 1. Statistical significance of correlates of Sr/Ca and Zn/Ca by sample.

Sr/Ca-Zn/Ca	r	p
Xoclán	0.218	0.246
PGJ	-0.531	0.114
Campeche	0.292	0.028
Xcambó	0.703	0.000005

Table 2. Average and standard deviation (σ) of values of Sr and Zn by sample. Sr average of the PGJ sample does not include outlier individual.

Sr	average	σ	Zn	average	σ
Xoclán	90.35	29.49	Xoclán	106.21	14.64
PGJ	68.57	37.38	PGJ	86.14	19.58
Xcambó	2113.00	715.60	Xcambó	77.06	20.98
Campeche	552.64	127.23	Campeche	85.32	15.10

229

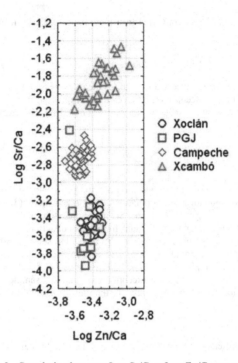

Figure 3. Correlation between Log Sr/Ca y Log Zn/Ca per sample.

DISCUSSION

We have noted during our study that Sr does indeed act as a dietary indicator by marking differences between populations, whereas Zn, an element under metabolic control [4] presents only minor variations within the total cohort of samples. In the following, we will discuss individually each of the trace element ranges obtained in our study.

Firstly, our results indicate higher values of Sr in both archaeological series (Figure 3). This result may be explained by the proximity to the coast of both archaeological samples. Authors like Lori Wright have reported that Sr levels tend to increase noticeable in samples from marine environments [9]. This would explain the origin of the discrepancies in strontium levels from Xcambó, when quantified, but in the case of Campeche then would not match the expected pattern. If we were to adhere to the classic biopurification model in the prediction of strontium element patterns, it follows that Xcambó´s population simply ingested a large proportion of vegetable foods in their diet compared to the other populations. But reality is not as simple as that, given the fact that Sr is a rather circumstantial addition to the crystalline structure of hydroxyapatite, as it is replaces haphazardly calcium within its crystalline molecule structure [8]. Another limitation is implied by the fact that the final strontium concentrations in bone tissue not only depend on the geological substrate but also on the quantity of calcium ingested dietary by the organism [10].

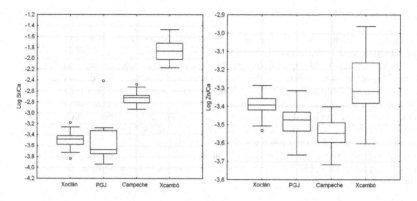

Figure 4. Boxplot showing ranges of Log Sr/Ca and Log Zn/Ca.

The results obtained in the forensic series pose a different caveat to be discussed here in view of the postindustrial quality of the diet it stands for. Modern foodstuffs are "delocalized" to a larger or lesser degree, as most of the food ingredients that make up urban diets, are not produced locally but have been shipped instead from many different geographic regions. This "globalized" provenience pattern has become widespread especially in large urban centers of the western hemisphere. This is thanks to the new infrastructure, which permits transport, storage and conservation of large quantities of foods.

The traditional biopurification model predicts that strontium element levels decrease when moving up the food chain, assuming a homogeneous geological setting. This assumption is invalid, naturally, when foods from different environments without known strontium levels make up the diet of a population. Especially post WWII western populations ingest a great proportion of highly processed and industrialized foods, most of which comes with supplements or conservatives unknown still in terms of Sr levels [11]. From this perspective, it is interesting to notice the similarities in Sr levels between the urban collection from the Xoclán cemetery and the PGJ series, with the noteworthy exception of one individual of the PGJ sample. This outlier sports exceptionally high Sr values (1037.31 ppm), which group it with the archaeological coastal concentrations measured in this study (Figure 3). Given the marine context in which this outlier was recovered from, we ask ourselves if the high Sr value was produced by a marine diet or instead by the diagenesis of the human remains within a burial environment laden with Sr.

Lastly, we intended to use the element Zn as a potential indicator of dietary protein consumption, although this assumption is highly questioned in the literature, where it is argued that Zn is controlled directly by the organic metabolism and is therefore regulated independently from dietary intake [3]. In our case we conclude that Zn levels are surprisingly similar between the forensic and colonial cohorts (Figure 3). Only Xcambó posesses a noticeable dispersion of Zn levels, which could be due to diagenetic contamination.

CONCLUSIONS

In summary, although we cannot easily observe dietary differences, Sr concentrations indicate a strong regional variation. On the other hand, due to the fact that diet is highly processed and import from very different geographic regions seems to be affecting Sr concentrations in bone by showing a very variable pattern. The present results confirm that Sr concentrations reflect in principle the geological substrate of the living spaces of the individuals under study. The two forensic series, albeit probably free of diagenetic contamination, do not match the expected profiles, an aspect which probably has to do with globalization and modern technology of food production and consumption. More aligned with geological factors are the Sr results of the two archaeological series. Except for the potentially altered sample from Xcambó, Zn concentrations appear more stable than the Sr levels, a tendency that we attribute to Zn being controlled by metabolism.

ACKNOWLEDGEMENTS

We thank M.C. Daniel Aguilar Treviño, M.C. Cinthia Mena Duran and in general, the staff of the CINVESTAV, Unidad Mérida, for kindly helping us during this investigation. We also thank indebted to the archaeological and forensic projects for letting us analyze bone samples. This work was partially supported by Conacyt Projects No. 123913 and FOMIX 108160 and 108528.

REFERENCES

1. D. Price, M.J. Schoeninger, G. Armelagos. *Journal of Human Evolution* 14 (1985) 419-447.
2. J. Burton. Bone Chemistry and Trace Elements Analysis, in: *Biological Anthropology of the Human Skeleton*. M. A. Katzenberg and S.R. Saunders. Eds. Wiley-Liss, New York, 2008, 443-460.
3. J. A. Ezzo. *American Antiquity*, 59 (1994) 606-621.
4. M. Rodríguez-Pérez. Living Conditions, Mortality, and Social Organization in Campeche during the Sixteenth and Seventeenth Centuries in *Natives, Europeans and Africans in Colonial Campeche: History and Archaeology*. V. Tiesler, P. Zabala, A. Cucina. Eds. University Press of Florida, Gainesville, 2010, 95-110.
5. S. Tejeda, V. Tiesler, W. Folan, M. Coyoc. Nutrición y estilo de vida en Campeche. in *Los investigadores de la Cultura Maya* 9. Tomo 2. Universidad Autónoma de Campeche. Campeche, 2001.
6. A. Cetina Bastida. *Población, nutrición y condiciones de vida en Xcambó, Yucatán*. B.A. Thesis. Facultad de Ciencias Antropológicas. Universidad Autonoma de Yucatán, Mérida, 2003.
7. Th. N. Sierra Sosa. *La arqueología de Xcambó, Yucatán, centro administrativo salinero y puerto comercial de importancia regional durante el clásico*. Doctoral dissertation. Facultad de Filosofia y Letras, Universidad Nacional Autónoma de México, México, 2004.
8. R. Lozano, J.P. Bernal. *Revista Mexicana de Ciencias Geológicas*, 22 (2005) 329-344.
9. L.E.Wright. *Diet, Health and Status Among the Passion Maya: a Reappraisal of the Collapse*. Vanderbilt University, Nashville, 2006.

10. J. H. Burton, T. D. Price. The Use and Abuse of Trace Elements for Paleodietary Research in *Biogeochemical Approaches to Paleodietary Analisis*. S. H. Ambrose, M. A. Katzenberg. Eds., Kluwer Academic/Plenum. New York, 2000, 159-171.
11. E.J. Moynahan, M.J. Jackson. Trace Elements in Man. Philosophical Transactions of the Royal Society. *Biological Sciences* 288 (1979) 65-79.

Mater. Res. Soc. Symp. Proc. Vol. 1374 © 2012 Materials Research Society
DOI: 10.1557/opl.2012.1393

Morphometric Characterization of the Maize: A Case Study in Postclassic Xaltocan

Naoli Victoria Lona[1]
[1]Posgrado en Antropología, Facultad de Filosofía y Letras - Instituto de Investigaciones Antropológicas, Universidad Nacional Autónoma de México (FFyL/IIA-UNAM), Circuito Exterior - Mario de la Cueva s/n, Ciudad Universitaria, Coyoacán, México, Distrito Federal, C.P. 04510, México. e-mail: naoliv@hotmail.com

ABSTRACT

Xaltocan used to be an island located North of the Basin of Mexico in the bed of the lake of the same name, the occupation of which has been continuous for 1100 years. The area's environment includes the island, the lake and the shore Xaltocan saline land and deep alluvial soil, the foothills and mountains, areas that provide different resources, whose exploitation is clear from the Paleoethnobotanical remains recovered from archaeological excavations.

Maize, basic resource in the daily life of Xaltocan, was also affected by sociopolitical and economic circumstances, so, it is possible that a different type of maize was used in each period of the locality. This is the interesting point to characterize morphometrically the maize from a context housing for the early Postclassic. To do so requires the application of different techniques (methodical process) to get to the characterization of corn: sample, flotation, separation and identification, measurement of corn through a microscope that could be complemented with electron microscopy scanning (SEM) to reveal possible microstructure.

In the measurement technique applied to the samples, the cob is divided into different sections; all of them can be measured, element by element, with a stereoscopic microscope for subsequent statistical analysis.

The study includes 231 sediment samples, distributed in 19 different depths and areas inside and outside of two dwelling domestic units. From 31,230 macrorremains, 6,140 were identified as *Zea mays*.

INTRODUCTION

Consumption of food is a biological need for all living beings; however, the human being has taken this activity to evolve into a biocultural need. "Evolution" is meant to be understood here as development without any "better" or "worse" connotation added to it. What this means is, it is known that through history, man fed on everything that nature offered him without "modifying" it , like fruits, plants and grains; or else, perhaps the dead animals he could claim. "Modifying" it is, in the sense of changing its natural cycle, intentionally promoting its reproduction.

Later on, man created tools and artifacts the uses of which he gradually specialized with the purpose of procuring food for himself. He developed his techniques to the grade of getting to be capable of "modifying" the natural cycle and to create a food supply by hunting, domesticating, fishing and cultivating.

Not contempt with his newly found ability to control his own food resources, he started to explore what we would consider today as cooking. In other words, food was no more directly consumed (that is, raw), but instead fire was applied to its preparation, hence giving birth to the crossing from a once purely biological need to a biocultural one. Together with the new ways to eat food, there also came new ways to store it, the most common of them being dry places like barns; stuffing, either using temperature or salt; low temperatures, among others.

Processes like the very simply described above, have implied constant organization and modification of the social-economical system both nowadays and back in the pre Hispanic epoch to be evidenced in different aspects: diet, artifacts, technology, clothing; that is, in every context of human life, both natural and cultural.

Xaltocan (Figure 1), the place studied in this research, is no exception; for in the pre Hispanic epoch, it had fishing, farming and salt regions due to the fact that it is an island located North of the basin of Mexico, on the bed of the lake of Mexico, which has been continuously occupied for 1100 years [1]. The area's environment (Figure 2), in general, and considering what's been said by Sanders, Parsons and Santley [2], includes the lake of Xaltocan and its saline shore, the deep soil of alluvium, foothills and the ridge, areas that provide different resources the exploitation of which can be deducted from the paleo-ethnobotanical remains recovered in the excavations.

In fact, the lake of Xaltocan was a source of edible animals, seasonal water birds, tule trees used to make baskets, carpets and roofs [3]. In addition, chinampa agriculture was developed there [5].

The deep soil of alluvium is located at the Eastern part of the lake, at the base of the mount of Chiconautla. Despite being a zone with scarcity of rain and strong frosts in the lower areas, intensive irrigation farming was practiced in the Postclassic period, providing diverse and varied products [3].

The low foothill (approximately 2260 to 2350 meters) presents a soil of 5 to 50 meters depth, but for the Postclassic period terracing was used, which could have provoked accumulation of soil and with it, its productivity. Original vegetation suggested is oak bush (*Quercus microphylla*) [2].

The middle foothill (approximately 2350 to 2500 meters), with moderate slopes and a soil that is similar to that of the low foothill, gets eroded with ease, but that is countered by terracing in order to retain both the soil and humidity, promoting agricultural activity. Original vegetation suggested is oak and xerophilous vegetation, various types of cactus, leguminous and grasses.

Regarding high foothill (approximately 2500 to 2700 meters), it is more prone to erosion due to slight soils, a bigger amount of rainfall and strong frosts, all of which promotes a very less productive farming. Original vegetation suggested is rainforest mixed with oak and pine.

The ridge (2700 meters over the level of the sea or more), more elevated areas of the Sierra de Guadalupe, is marginal to agricultural activiy, but it represents an important source of resources useful for building, fuel and animal supply. Original vegetation suggested is oak and pine forest [3].

To this day and in general, the identified macrobotanical remains are mainly plants adaptable or resistant to saline soils like the genres *Trianthema, Portulaca, Chenopodium, Atriplex, Suaeda* [7]; taxa that is typical of humid soils such as Polygonaceae [7], superficial water and swamped areas such as Cyperaceae [7]; other genres like *Amaranthus, Physalis, Solanum, Salvia, Phaseolus,* varios cactaceae (*Opuntia, Myrtillocactus geometrizans*) that include species both edible and collected from secondary instances of wild vegetation, as well as domestic cultigens like *Zea mays* [3].

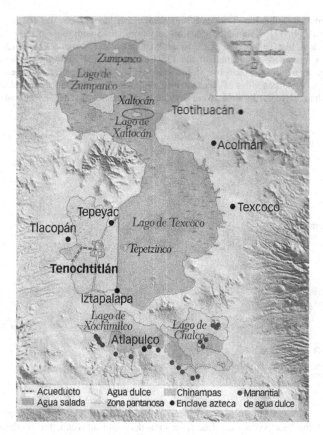

Figure 1. Xaltocan is located north of the Basin of Mexico; UTM coordinates are: 14Q 469379E, 21817549N; to 2,556 masl. [6].

Such exploitation of resources responds to economic and political interests from the societies that inhabit that type of lake environments, just like Xaltocan which went from first cultivating maize for local consumption to having to modify its production after being turned into a tributary people of the Triple Alliance [1], giving maize as a tribute product. That is, the resource used and/or controlled by a cultural group, is then affected depending on the sociopolitical situation, and therefore also on the economical situation that group experiences in a certain period of time.

According to historic sources, Xaltocan was funded in the 11th Century A.D., immediately after the fall of Tollan [8, 9]. During the 12th and 13th Centuries, it was developed as an important regional center (1200-1500 A.D). It probably was the capital of Otomi spkeaking peoples in the South of today's State of Hidalgo and the North of the basin of Mexico [8, 10]. The Vaticano-

237

Rios codex mentions Culhuacan, Tenayuca, and Xaltocan as having been the dominant powers before the Mexica [10-11]. Xaltocan's heads-of-state were allied with those of other important centers of the basin, like Tollan, Tenayuca, Huexotla, Chalco and Azcapotzalco [8,12-13]. This organization was modified around half of the 13th Century because Xaltocan sustained a long lasting war with its neighbor Cuautitlan; in the year of 1395 A.D., the latter defeated Xaltocan with help from its Tepaneca allies of Azcapotzalco. Xaltocan was abandoned and remained like that for more than 30 years. Its lands were divided by the heads-of-state of Azcapotzalco and Texcoco [8-9].

Geoenvironmental system
Model: Berenice Solís Castillo

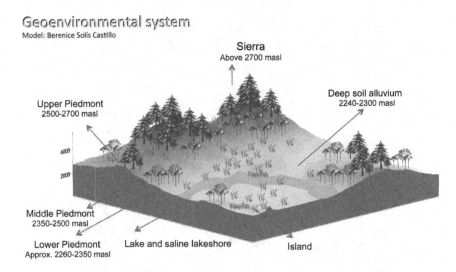

Figure 2. The environment of the area in general and considering Sanders, Parsons and Santley (1979) includes the lake Xaltocan, saline lakeshore, deep soil alluvium, the foothills (lower, middle and upper Piedmont) and the mountains (Sierra) [3].

In 1428 A.D., Xaltocan fell under dominance of the Triple Alliance of the Aztecs. The heads-of-state of Tenochtitlan and Tlatelolco settled tribute-paying farmers in Xaltocan [32], and a military head of state (*cuauhtlatoani*) was sent from Tenochtitlan [12]. Under the Triple Alliance, Xaltocan paid tribute to both Texcoco and Tenochtitlan [8, 12]. In 1521 A.D., Xaltocan was attacked and set on fire by Hernán Cortés [14], so marking the end of the pre-Hispanic epoch. Díaz del Castillo mentions that Xaltocan was a very rich people, and very loyal to Tenochtitlan [15]; for he recounts that decorative artifacts of the image of Huitzilopochtli belonging to the Tenochca high priest of Tenochtitlan were sent to Xaltocan to be safeguarded [18].

THEORY

The way in which society develops depends on a number of variables – biological, economical and political, as well as those determined by the mere preference or particular benefit of the person or group in charge. Therefore, the economic change for the pre Hispanic epoch can also be evident, not only in the products *per se*, but in the population an its day-to-day activities of maintenance.

"Historically speaking, a sociopolitical unit is defined according to its size, population density, economy and the political relationships existing between one unit and the others." [19] Every place starts with a period of adaptation to the environment, both natural and cultural. As its settling process progresses, strategies for the establishment of sociopolitical, economic and cultural relationships with others are created, defined, modified or adopted, all of which reflects in the material evidence registered in different contexts.

Table I. Historical sources [1]: Alva Ixtlixóchitl [8, I: 321-323, II: 89-90]; *Anales de Cuauhtitlan* [9], Carrasco [10, 260-261]; Nazareo [12, 109-129].

Chronology of events in Xaltocan's history, according to chronicles

Approximate Years A.C.	Event
1200 - 1250	Xaltocan is capital of the Otomí nation
	Major political allies are Culhuacan and Tenayuca
1250	War begins between Xaltocan and Cuauhtitlan (part of the Tepaneca empire led by Azcapotzalco)
	Xaltocan is conquered by the Tepaneca empire led by Azcapotzalco, assisted by Cuauhtitlan and Tenochtitlan.
1395	Xaltocan's nobles flee and Xaltocan is abandoned as a city-state center for three to four decades
	Lands in Xaltocan are divided among rulers of Azcapotzalco and Texcoco
1420-1428	The empire of Azcapotzalco is attacked by the Triple Alliance of Tenochtitlan, Texcoco and Tlacopan, and is defeated
1435	Xaltocan is resettled by Tenochca
1430-1521	Xaltocan is a dependency of Tenochtitlan, with a Tenochca governor
(ca.)	Xaltocan is also a tributary province of Texcoco, with an acolhua calpixque or tribute collector assigned to it

As population increases, in case its development favors such increase, economy also has its own process. Such process would be determined by the use of resources which at the same time is determined by local and regional variables [1]. Julian Steward sustains that the former includes natural resources, technology and labour [21]; as an imperial system, which is intended to transcend the limits of its localities [1], force or the threat of it is applied to obtain products and work from subordinated populations in a certain geographic area [22-26], thus affecting local production and the place's economy.

Through the Postclassic period, Xaltocan created, adapted, modified, discarded and adopted new strategies of subsistence that were directly related to its environment and also to the historic events that took place: formation, peak, and subjugation by the Triple Alliance.

Such circumstances can be made evident in the material culture resulted from the technological production: building systems, production and consumption of rock and ceramic (styles that are characteristic of that time, such as the Aztec 1, II, III, and IV) or agriculture (by terracing or chinampa, irrigation or seasonal; off of those contexts macrobotanical remains were recovered), all of which would mean a background of local consumption (diet) and sociopolitical relations (trade).

Maize, fundamental resource in the daily life of Xaltocan, was also affected by such sociopolitical and economic circumstances, making evident a differential diversity in each time period of the locality. According with Ch. Morehart, variation in the types of *Zea mays* changed according to the sociopolitical and economic circumstance occurring in Xaltocan [27].

Phase 1

In phase 1 (900 to 1100 A.D., early Postclassic period): establishment of Xaltocan; great variability of *Zea mays* as there was not a well-established farming system, nor there was a political obligation towards other entities, only local needs were to be covered.

Phase 2 and 3

In Phase 2 and 3 (1100 to 1300; 1300 to 1400 a. C., middle Postclassic period): the power of Xaltocan increases, a lesser variability, possibly due to the market relations and tribute controlled by Xaltocan as it was the power center after establishing chinampa farming as a way to systematize farming and adapt it to the environment. Use of resources in the basin of Mexico during the middle and late Postclassic period, was altered by changes in the tributing and marketing systems, and by changes in the regional population [1]. As Xaltocan settled as an important power center, it established its systems – economic, political and social: it managed the lake resources, both animal and vegetal, the latter including maize; Xaltocan strengthened its political organization based on the marital alliances with its regional neighbors, and designated activities within the Xaltocamecan society where the inhabitants of domestic units were self-sustainable regarding the available resources [1].

Before being subjugated by the Triple Alliance, Xaltocan managed a generalized and heterogeneous economy [28], where a small range and volume of early trade, judging by the existence of residencies of local political elites, long distance exchanges and specializes craftsmen; besides, there was a lack of pressure on the productive efficiency due to the low population density and low levels of tribute and of market competition [1].

Phase 4

In Phase 4 (1431 to 1521 A.D.; late Postclassic period): conquest of Xaltocan; great variability as Xaltocan lost its hegemony and another sociopolitical model was introduced, that of the Triple Alliance; that could've provoked some adaptations in the life style of the locals, including diet. By the late Postclassic period, the basin of Mexico was politically consolidated through a network of paths and channels that encouraged concentrations both of people and of wealth, which at the same tame gave way to rural-to-urban exchange centered on the Alliance's most important capital, Tenochtitlan. Under this regime, product mobility increased by the use of the market system [1].

Xaltocan's autonomy and prosperity during phase 3 of occupation (1300 to 1430 A.D.) is very clear when compared to phase 4 (1430 to 1521 A.D.), in the time when the Triple Alliance dominated the area [1]. Once the Triple Alliance subjugated Xaltocan, economic and political organization of the latter was altered, for not only a Tenochca head of state was imposed to it, but also maize was demanded from it as a tribute for Tenochtitlan and Texcoco. This prevented redistribution and management of resources, because the chinampas that originally were used to produce maize exclusively for local consumption possibly started to be utilized to produce also for tribute payment [1].

"Xaltocan must have been totally absorbed within the regional market system. It must have constrained its range of productive activities and concentrated its production on food for sales in Tenochtitlan. Although Xaltocan only had a limited amount of farming lands, it was able to exploit the abundant lake resources, which included protein-rich food sources like fish, water birds, insects and alga. These food sources must have had great demand as supplements of maize in the urban diet" [1].

Off the formerly presented data, an interest emerged to explore how production of *Zea mays* could have varied under these new political and economic circumstances, as Xaltocan was shaped in domestic contexts; and how much variability in the types of maize was registered, regarding both quantity and quality, so an study of the botanical material obtained of excavations will have to be done, so it can be compared to that of the farming contexts and basis can be set for complementary information about if what was produced was just the same as it was consumed.

Thus, the purpose of this investigation is to establish the variability of maize from domestic backgrounds of the phase occupation 1, corresponding to the early Postclassic period (900 to 1100 A.D.) using the morphometric characterization of entire elements of *Zea mays* by measuring of different sections of the elements of maize in order to determine variability based on ranges established by statistical programs. For the study of maize, corncob has been divided in different sections: olote (cob), cupule, glumas, tricomas, grains, leaves, stems, among others [27,30]; all of those can denote different varieties according to its micro and macrostructure. The latter can be characterized by the morphometric measuring of this inner structure, with the electronic microscope or even DNA analysis.

This study includes grains, cupules (a section of the corncob where grains are placed), and olotes (the coral section of the corncob, with a cylindrical shape, were cupules are placed), for its further comparison with material registered in farming context (chinampa) from the same time period analyzed by Ch. Morehart & E. Brumfiel [1, 27]. The materials studied come from domestic and farming contexts (chinampas) of the island of Xaltocan for the early Postclassic period registered in the research of the archaeological project named "Strategies in the Domestic Units of Postclassic Xaltocan, Mexico" coordinated by E.M. Brumfiel, developed from 1987 to date.

METHODOLOGY

Laboratory analysis consists in the floating of sediments, weigh-in of the floated remains and its sieving, as well as the work of each element on the stereoscopic microscope: separation of macro remains (classified by botanical, bone, shell, wood, carbon; as well as carbonized, mineralized or naturally preserved), taxonomic identification, measuring of each element entire (with the use of a peephole calibrated in millimeters) and accounting of all the total macro remains (Figure 3). Such data is registered in the laboratory format, which was filled at hand as analysis is made. The macro remains are safeguarded in pristine gel capsules with its nomenclature of identification including project and context (Figure 4).

241

Once general analysis was done, we proceeded to elaborate the data base, which consisted in the design of it as well as the entering of the data (Dbase, Access and Excel). Having already entered the materials, *Zea mays* were separated by samples, thus creating a different data base with information only regarding to maize. From the latter, and against the study materials, grains, cupules, olotes and a number of other fragments [27]; classification of the elements is further made by complete and incomplete ones, for only the former ones are measured (Figure 3 and 4).

Table II. Model for maize in Xaltocan: socio-economic and political changes are evident in the resources that produce and consume the inhabitants of the island [1, 4, 27, 32].

Occupation Phases of the Postclassic Xaltocan (Millhauser, 2005)				Context	*Zea mays*
Phase	Year CE	Ceramic type	Time	Sociopolitical situation	Morehart & Eisenberg Model (2009)
1	900 - 1100	Aztec I	Early Postclassic	Settlement	High variability of the maize (shorter cupules)
2	1100 - 1300	Aztec I-II	Middle Postclassic	Settlement established as a center of power: Xaltocan as a capital Otomi, political allies Tenayuca and Culhuacan	Low variability of the maize (longer cupules)
3	1300 - 1430	Aztec II-III	Middle - Late Postclassic	Begins war with Cuauhtitlán; Xaltocan is conquered by the empire Tepaneca; xaltocameca nobility flees, Xaltocan is abandoned as the capital city-state for 3 or 4 decades and divided between Azcapotzalco and Texcoco; Azcapotzalco is conquered by the Triple Alliance (Tenochtitlan, Texcoco and Tlacopan)	Low variability of the maize (longer cupules)
4	1430 - 1521	Aztec III	Late Postclassic	The Tenochca populate Xaltocan once again, a Tenochca goverment is imposed and Xaltocan is becomes tributary province of Texcoco administered by a Acolhua calpixque	High variability of the maize (shorter cupules)

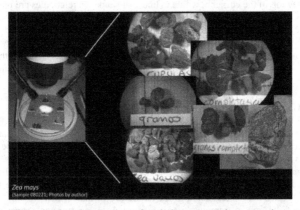

Figure 3. Identification of maize sections for their conservation: complete and incomplete cupules, complete and incomplete grains, incomplete olotes and various other fragments.

The methodology for the study of macro botanical remains starts with the take of sample *in situ*; floating technique is applied to such samples in order to separate macro remains, which are further classified by type with the use of a stereoscopic microscope: carbon, seed, bone, leaves, and at the same time if they're carbonized, mineralized or none of the two; once the seeds are separated, they're classified by family, genre and species; measuring is made to every complete seed as well as the accounting of the number of individuals in each set of samples. As maize is grouped in this manner, it is separated by complete and incomplete elements in order to base the analysis only in the former ones, for they're the ones that permit the sample to be standardized (or not). Everything is then entered in data basis.

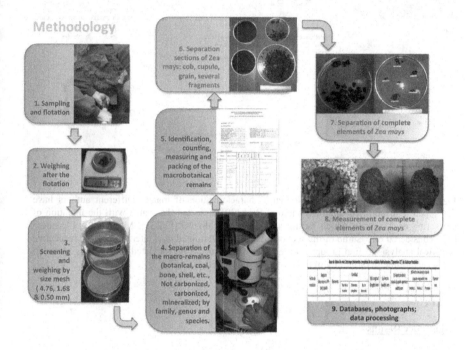

Figure 4. Methodology for studies of macro botanical remains.

The measuring technique applied to the samples was created by R. Bird [30] and later modified by Ch.Morehart & C. T. Eisenberg [27], it consists in the element-by-element measure to a scale of millimeters using a stereoscopic microscope, of: length, width and thickness of the grains; length, width, thickness and aperture of the cupules and the width and wing width of the cupules; and diameter of the raquis of the olote in the case it is available, if it's not then only cupules are measured individually, even if they're adhered to the incomplete olote (see Figures 5 and 6).

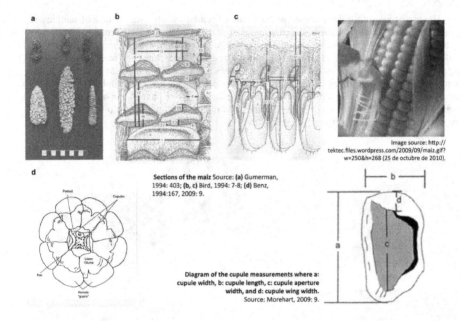

a b c

Image source: http://
tektec.files.wordpress.com/2009/09/maiz.gif?
w=250&h=268 (25 de octubre de 2010).

d

Sections of the maiz Source: (a) Gumerman,
1994: 403; (b, c) Bird, 1994: 7-8; (d) Benz,
1994:167, 2009: 9.

Diagram of the cupule measurements where a:
cupule width, b: cupule length, c: cupule aperture
width, and d: cupule wing width.
Source: Morehart, 2009: 9.

Figure 5. Methodology for morphometric characterization of maize. Different authors have developed and modified this methodology Bird [30] dissected the maize with the intention of measuring each area in which the macro remain was divided [28-31]. Morehart & Eisenberg [27], considered only some of the sections that Bird determined, and later applied the respective statistical analysis to find groups, in the cases where such groups existed in the study material, and so determine if variability of the sample is high, average or low.

RESULTS

Laboratory Results

The study includes 231 samples of sediments from "Operation Z3". "Operation Z3" is a part of the archaeological project "Strategies of the Domestic Units in Postclassic Xaltocan, Mexico", directed by PhD. Elizabeth M. Brumfiel, in its season of 2008. The samples were distributed in 19 different levels of depth and areas both inside and outside of two housing units. 31,230 macro remains were obtained from those samples, where 16,074 correspond to carbon fragments and 15,156 to botanical remains.

Zea mays: Cupule
(Photo by author)

Diagram of the cupule measurements where
a: cupule width, b: cupule length, c: cupule
aperture width, and d: cupule wing width.
Source: Morehart, 2009: 9.

Stereomicroscope
12-60x (Wilde M3)

Zea mays: Cob
fragment
where the
cupules are
measured
(Photo by
author)

Figure 6. The study sample is conformed in its majority by cupules, whether they are off the olote or still adhered to the corncob; if the latter is the case, measurements of the cupules are made individually without considering the measurement of the olote as it is incomplete.

Of the latter, 6,140 were identified as *Zea mays* distributed in 198 samples, from the latter 3,418 elements were registered and analyzed: 3,079 cupules, 214 grains and 125 olote fragments. Out of all the *Zea mays,* 2,722 fragments of maize were accounted, material that was not considered for the analysis. Thus, complete elements were considered for the next phase of the study (1309 cupules). Once they were measured and entered in the data base, we proceeded to the statistical analysis to produce a reliable and comparable sample between phases of occupation.

Statistics Analysis

In order to get to establish the type of analysis to be applied, the number of cases included in the sample has to be considered. Thus, we proceeded to design a distribution diagram of it, taking on account only the complete elements to take measurements that will serve to do the morphometric characterization of the study material, just as it was mentioned before. In the case of the grains: length, width and thickness; and as for the cupules: length, width, cupules aperture and wing width 1 of the cupules (Figure 5). The wing width of the cupules can be taken from both extremes, the wider of the two being name as "1", and the slighter as "2". For the study, only "1" was considered.

By the use of Statistical Package for the Social Sciences (SPSS), the distribution diagram of the sample was built quantifying not only how many cases there are, but also their type (grain, cupule, olote (cob)), how many of them there are by phase of occupation (phase 1, phase 2, colonial/historical, historical or non-defined) and their location (North house or South house) (Table III).

Table III. Distribution of the sample according to type of element (grain, cupule, olote (cob)), phase of occupation (phase 1, 2, Colonial/Historic, Historic and non-defined context) and location of the elements (North house or South house).

Occupation Phase	House location: north/suoth	Item type			Total by Occupation Phases
		Grain (complete / incomplete)	Complete Cupule & cupule on incomplete cob	Cob (fragment)	
Phase 1	North (F7)	39	762	13	814
	South (F29)	64	491	10	567
	Subtotal	103	1253	23	**1381**
Phase 2	North (F7)	0	0	0	0
	South (F29)	4	56	1	61
	Subtotal	4	56	1	**61**
Colonial / Historic	North (F7)	0	0	0	0
	South (F29)	0	3	0	3
	Subtotal	0	3	0	**3**
Historic	North (F7)	0	3	0	3
	South (F29)	0	0	0	0
	Subtotal	0	3	0	**3**
Undefined context	North (F7)	0	0	0	0
	South (F29)	5	62	1	68
	Subtotal	5	62	1	**68**
TOTAL	North (F7)	39	765	13	817
	South (F29)	73	612	12	699
	Total	112	1377	25	**1516**

These data are greatly useful, because they permit to know what type of element (grain, cupules) can be comparable regarding the phase of occupation and location of the sample (North or South house) by using descriptive statistics (univariated), to further determine what analysis to apply and to what type of sample, regarding the multivariated statistics (main components, conglomerates) and determine maize variability (whether it's high or low).

This way is made evident that there is a higher concentration of cupules for both phases 1 and 2. Phase 1 is only considered because that's the one that accounts for a reliable sample for the multivariated statistical analyses in benefit of the characterization of the material (1,253 elements). Despite the fact that the sample of the phase 2 (56 elements) is smaller than that of the phase 1, the amount of elements permits a comparison of the cupules of each phase and demonstrate its variability based on its morphometric measurement (a total 1309 cupules of maize), being the actual sample of the study after analyzing 3,418 total maize elements from a major sample of 31,230 macro remains (see Table IV).

Table IV. Descriptive Statistics with Statistical Package for the Social Sciences (SPSS).

Occupation Phase	House location: north/suoth	Item type			Total by Occupation Phases
		Grain (complete / incomplete)	Cupule & cupule on incomplete cob	Cob (fragment)	
Phase 1	North (F7)	39	762	13	814
	South (F29)	64	491	10	567
	Subtotal	103	1253	23	**1381**
Phase 2	North (F7)	0	0	0	0
	South (F29)	4	56	1	61
	Subtotal	4	56	1	**61**
				Total	**1442**

In order to have reliability in the sample, T-tests were applied, where only when they give a result of "$p<0.05$" can the difference between occupation phases actually be considered, having a measurement of <0.05 for all the measurements of the cupules considered (length, width, aperture of the cupules and width 1 of the cupules' wing); therefore, the difference is considered reliable [16-17, 20, 33].

Comparison Confidence Interval Graphs, to 95% Confidence Interval (CI), between phases 1 and 2 thus can be done, considering the length, width, aperture of the cupules and width 1 of the cupules' wing. A clear difference between maize of both phases was observed, the size of the maize of the phase 1 being significantly superior (numeric index that marks the little central circle, which represents the sample's arithmetic media (X), Figure 7).

As of the length, width, aperture of the cupules and wing width 1 of the cupules, a minor variability of it is made evident in phase 1 with respect to phase 2 (length of the line in the graphic represents the variability of each sample (CI), Figure 7).

A second exercise was conducted, grouping the cupules by context of the phase instead of doing it by phase of occupation; in other words, phase 1 maize recovered from the housing unit called North and South. It was observed that the maize from the North house is significantly bigger than that of the South house; and the variability regarding length, width, cupules aperture and wing width 1 of the cupules is very similar (Figure 8).

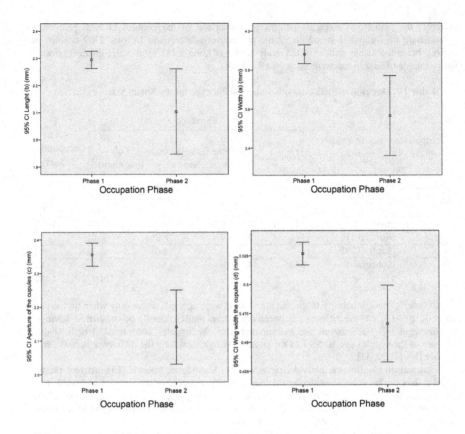

Figure 7. Only when the result of the "T" test is "$p < 0.05$" can we actually consider the difference between phases of occupation. As the graphic shows, maize of phase 1 is less variable and of a bigger size than that of phase 2, which is more variable and smaller.

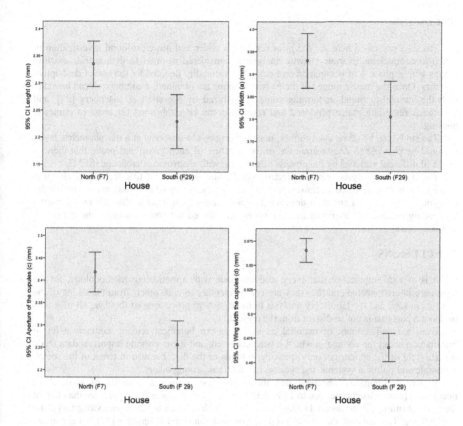

Figure 8. Only when "T" test's result is "p< 0.05" can we actually consider the difference between cupules of the phase 1 of occupation; such is the case when maize diversity in both houses is similar: the maize from the North house is considerably bigger than that of the South house.

DISCUSSION

The data presented here are the prior result of a wider and more profound investigation in which to conduct multivariate statistic analysis is considered in order to demonstrate existing groups within phase 1 of occupation and detail the variability denoted by the use of descriptive statistics. Once the metric ranges of the possible groups are obtained, a comparison can be made with the variability found in farming contexts analyzed by Morehart & Eisenberg [27], and hypothesis regarding maize produced and consumed can be established for areas of domestic housing.

The study can be taken even further; once the groups are observed at a morphometric level, it would be possible to demonstrate the micro structure of each group and prove that there is maize of different varieties by the process of scanning with electronic microscope (SEM).

This type of analysis requires electron conducting samples, so this technique, although effective, it is also destructive; however, in the case of the maize of the study it is ideal, for the elements are carbonized and their destruction is not necessary, instead of that it is only required the making of tests to observe if that carbonization allowed the conservation of the material's microstructure.

CONCLUSIONS

It is of vital importance that every study is made with a meticulous methodology, for the lapses are interconnected and thus they are complementary to each other; from the taking of the samples *in situ*, and the laboratory analysis to the statistical processing of the data, all of it tells us about a past that is not so distant from the present.

Even a small amount of material as it is a macro botanical remain, contains a lot of information about the context in which it was recovered, and it can provide important data about the life style of the epoch; not only speaking in terms of the diet, but also in terms of the social, economic and political systems, the systems of traditions, among others.

It was preliminarily observed that the maize from domestic contexts of phase 1 of occupation (early Postclassic: 900 to 1100 A.D.) has less variability and a bigger size than that of the phase 2 (middle Postclassic: 110 to 1300 A.D.), which resulted to have more variability but a smaller size. Considering the model proposed by Morehart and Eisenberg [27], where maize from phase 1 coming from farming contexts (chinampa) has high variability and is smaller than that of phase 2, which registered a lesser variability and a bigger size; a clear difference can be observed between both results, as they are completely opposite.

The former argument could be related to the fact that the maize registered for the chinampas was destined to a different utilization or group of consumption, and that the maize from the domestic units studied came from somewhere else, like the head of the market organization, exchange or any other; however, what is important is to remark that, while being an island and being both contexts so close to each other, they do not portray the same material, as it could be expected. A clarification of all this is intended to be made in the future, once all the analyses are applied, so it is possible to prove what descriptive statistics is providing us so far.

ACKNOWLEDGMENTS

We acknowledge the support of Dra. Emily McClung, Diana Martínez, Emilio Ibarra, Cristina Adriano from LPP-IIA-UNAM; CONACyT; Dra. Elizabeth Brumfiel, Christopher Morehart, Dan Eisenberg, Kristin de Lucia and Flor María Rivas from Xaltocan Project; Dra. Annick Dannels, Gerardo Jiménez Delgado from Posgrado en Antropología FFyL/IIA-UNAM, Dr. José Luis Castrejón from Escuela Nacional de Antropología e Historia (ENAH); José Luis Hernández Jiménez from Instituto de Investigaciones Antropológicas de la UNAM; Dr. Jorge Gama, Dra. Carolina Jasso and Dra. Adela Margarita Reyes Salas from Instituto de Geología, UNAM; Dr. José Luis Ruvalcaba from International Material Research Congress, Berenice Solís Castillo, Jocelyn Alcántara, Daniela Huber, Alejandra García, Violeta Corona, Erick Macías, Dra. Angelina Barreiro, Dra. Verónica García and Mtro. Adrián Benítez Ortega.

REFERENCES

1. E.M. Brumfiel, *La producción local y el poder en el Xaltocan Posclásico / Production and Power at Postclassic Xaltocan*, Instituto Nacional de Antropologia e Historia - University of Pittsburgh, Mexico, 2005.
2. W.T. Sanders, J.R. Parsons, R.S. Santley, *The Basin of Mexico: Ecological Processes in the Evolution of a Civilization*, Academic Press, Nueva York, 1979.
3. Emily McClung de Tapia, D. Martínez Yrizar, Paleoethnobotanical Evidence from Postclassic Xaltocan, in: *La producción local y el poder en el Xaltocan Posclásico / Production and Power at Posclassic Xaltocan*, Instituto Nacional de Antropologia e Historia - University of Pittsburgh, Mexico, 2005, 207-232.
4. J. Millhauser, Classic and Postclassic Chiped Stone at Xaltocan, in *La producción local y el poder en el Xaltocan Posclásico / Production and Power at Posclassic Xaltocan*. Instituto Nacional de Antropologia e Historia - University of Pittsburgh, Mexico, 2005, 267-317.
5. C.D. Frederick, B. Winsborough, V.S. Popper, Geoarchaeological Investigations in the Northern Basin of Mexico / Investigaciones geoarqueológicas del norte de la Cuenca de México, in: *La producción local y el poder en el Xaltocan Posclásico / Production and Power at Posclassic Xaltocan*. Instituto Nacional de Antropologia e Historia - University of Pittsburgh, Mexico, 2005, 71-117.
6. http://i614.photobucket.com/albums/tt223/regiosfera/mapa.jpg (consulted in November 10th, 2009; modified by author).
7. J. Rzedowski, G. Calderón de Rzedowski (eds.), *Flora fanerogámica del Valle de México*, vol. I, México. Limusa, Mexico, 1979.
8. Fernando Alva Ixtlilxóchitl, *Obras Históricas*, E. O'Gorman (ed.), Universidad Nacional Autónoma de México, Mexico, 1975.
9. Anales de Cuauhtitlán, Códice Chimalpopoca, P.F. Velázquez (trad.), Universidad Nacional Autónoma de México, Mexico, 1945 [1885], 1-118.
10. P. Carrasco, *Los otomíes: cultura e historia prehispánica de los pueblos mesoamericanos de habla otomiana*, Biblioteca Enciclopédica del Estado de México, Mexico, 1950.
11. N. Davies, *The Toltec Heritage: From the Fall of Tula to the Rise of Tenochtitlan*, Civilization of the American Indian series, Vol. 153, University of Oklahoma Press, Norman, 1980.

12. P. Nazareo de Xaltocan, Carta al rey don Felipe II, in:, *Epistolario de Nueva España*, vol. 10, F. del Paso y Troncoso (ed.), Antigua Librería Robredo, Mexico, 1940, 109-129.
13. Anales de Tlatelolco, S. Toscano, H. Berlin, y R. H. Barlow (eds.), Antigua Librería de José Porrúa e Hijos. México, 1948
14. H. Cortés, *Cartas de Relación*, Porrúa, Mexico, 1970 [1960].
15. B. Díaz del Castillo, *The Discovery and Conquest of Mexico*, A. P. Maudslay (trad.), Nueva York, Noonday Press, 1956 [1632].
16. R.D. Drennan, *Statistics for Archaeologist. A Commonsense Approach*, Plenum Press, New York and London, 1996.
17. M. Ferrán Aranaz, *SPSS para Windows. Análisis Estadístico*, Osborne/ McGraw-Hill, España, 2001.
18. M. León-Portilla, *The Broken Spears: The Aztec Account of the Conquest of Mexico*, Beacon, Boston, 1962.
19. R. García Chávez, *El altepetl como formación sociopolítica de la cuenca de México. Su origen y desarrollo durante el posclásico medio*, (consulted on October, 2010) in: http://www.ucm.es/info/arqueoweb/numero8_2/garciachavez.htm
20. S. Shennan, *Arqueología Cuantitativa*, Crítica, Barcelona, 1992.
21. J.H. Steward, *Theory of Culture Change*, University of Illinois Press, Urbana, 1955.
22. E.R, Wolf, *Europe and the People without History*, University of California Press, Berkeley, 1982.
23. R. Hassig, *Aztec Warfare: Imperial Expansion and Political Control*, University of Oklahoma Press, Norman, 1988.
24. T.N. D'Altroy, *Provincial Power in the Inca Empire*, Smithsonian Institution Press, Washington, D.C., 1992.
25. C.M. Sinopoli, *Annual Review of Anthropology* 23 (1994) 159.
26. S.E. Alock, T.N. D'Altroy, K.D. Morrison, C.M. Sinopoli (eds.) *Empires*, Cambridge University Press, Cambridge, 2001.
27. Ch. Morehart, T.A. Eisenberg, *Journal of Anthropological Archaeology*, 29 (2010) 94-112.
28. G. Stein, Segmentary States and Organizational Variation in Early Complex Societies: A Rural Perspective, in: *Archaeological Views From the Countryside*, G.M. Schwartz y S.E. Falconer (eds.), Smithsonian Institution Press, Washington, D.C., 1994, 10.
29. B.F. Benz, Reconstructing the Racial Phylogeny of Mexican Maize: Where Do We Stand?, in: *Corn & Culture in the Prehispanic New World*, S.Johannessen & Ch.A. Hastorf (eds.), Westview Press Inc., Boulder, 1994, 157-180.
30. R. Bird, Manual for the Measurement of Maize Cobs, in: *Corn & Culture in the Prehispanic New World*, S.Johannessen & Ch.A. Hastorf (eds.), Westview Press Inc., Boulder, 1994, 5-22.
31. G. Gumerman IV, Corn for the Dead: The Significance of Zea mays in Moche Burial Offerings, in: *Corn & Culture in the Prehispanic New World*, S.Johannessen & Ch.A. Hastorf (eds.), Westview Press Inc., Boulder, 1994, 399-410.
32. F. Hicks, Xaltocan under Mexica Domination 1435-1520, in *Caciques and Their People: A Volume in Honor of Ronald Spores, Ann Arbor*, Museum of Anthropology, The University of Michigan, Anthropological Papers 89, 1994, 67-85.
33. PhD José Luis Castrejón, personal communication, 2011.

Mater. Res. Soc. Symp. Proc. Vol. 1374 © 2012 Materials Research Society
DOI: 10.1557/opl.2012.1394

Self Sacrifice Awls in Cantona, Puebla, Mexico

N. Valentín Maldonado[1], G. Pérez Roldán[2]

[1]Subdirección de Laboratorios y Apoyo Académico, Instituto Nacional de Antropologia e Historia (INAH). Moneda 16, Centro, México, DF, CP. 06060, Mexico.
e-mail: nvalentinm@hotmail.com

[2]Universidad Autónoma de San Luis Potosí, Avenida Industrias 101-A, Col. Fraccionamiento Talleres, San Luís Potosí, CP.78494, Mexico. e-mail: gilgertions@yahoo.com.mx

ABSTRACT

In the ancient city of Cantona (600 B.C a 1050 A.D.) in the region of Puebla, Mexico, a great number of awls made of animal bones were found inside several offerings and burials. The present paper presents the identification of species used to elaborate these awls, along with the formal characteristics of the objects and their manufacturing techniques. The last was studied through experimental archaeology and the analysis of modified traces by Scanning Electron Microscopy (SEM).

INTRODUCTION

Cantona is an archaeological site located at the far east of the Central High Plateau in the north-central section of the Puebla eastern basin. It was an important settlement from 600 B.C. to 1050 A.D. [1] (figure 1), due to its privileged geographical localization; this site connected the communities of the central high plateau with the Mexican Gulf coast. The Cantona Archaeological Project was started in 1992 with the direction of professor Angel García Cook, through this project several materials have been recovered, among them some animal bone remains from three excavation seasons. Until now, 5000 animal bones (approximately) have been identified. The analyzed bones belong to three vertebrate groups: reptiles, birds and mammals. The more numerous of them were those of mammals, which the more abundant are the Cervidae family with the 99.5% with three genres and four species, being the white tailed deer (*Odocoileus virginianus)* the most present within the site.

From the modified animal and human bone materials, worked objects were recovered, such as straighteners, polishers, decorators, burnishers, knives and awls, among others [2].

For this paper, we selected a group of awls with active cutting edges whose raw materials consisted of bones from two large Mexican carnivores of great ritual relevance.

The main objectives of this research are the identification of the animal species used in Cantona, as well as establishing the function of the cutting awls and identify the manufacturing techniques of them.

MATERIALS AND METHODS

As we said before, the osseous material came from the Cantona archaeological site and was sent by the professor Angel García Cook to the Archaeozoology Laboratory "M en C. Ticul Álvarez Solórzano" of the Subdepartment of Laboratories and Academic Support, INAH, for its analysis.

The identification was mainly made by direct comparison with specimens from the osteological reference collection of the same laboratory, and for the taxonomic classification and geographic distribution we use the one given by Wilson and Reeder [3]. In a first step optical microscopy was carried out at 10X, 30X and 63X.

Scanning Electron Microscopy (SEM), was perfomed at 100X, 300X, 600X, and 1000X (Figure 8b). The last one was used with the same parameters proposed by Velázquez Castro [7]: High Vacuum Mode (HV), 20 kV of energy, SEI signal, spotsize of 42 and 10mm of work distance.

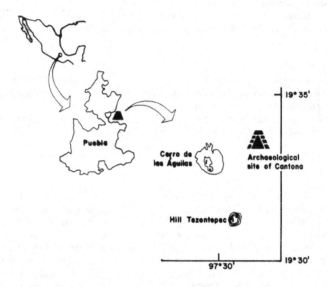

Figure 1. Location of Cantona archaeological site in Puebla, Mexico

BONE IDENTIFICATION

Nine osseous remains from mammals of carnivore order and Canidae and Felidae families were analyzed, its characteristics are stated below:

Family Canidae

Canis lupus baileyi.- This specie is commonly known as gray wolf, is one of the large mexican carnivorous. *Canis lupus* abundant over much of the North Hemisphere has several subspecies reported throughout America, but the one identified for Mexico is *Canis lupus baileyi* which could be found from the North of the country, passing through the Central High Plateau to Oaxaca. The gray wolf was much venerated in prehispanic Mexico and it has been found in several offerings and burials of different archaeological sites, such as Cantona.

254

The identified remains of this species were three; one metatarsus III, one phalanx and two that form the third toe of the right back leg, these were found at CJP5-1 (ball game complex), top, N 5.98 to 6.14, E 6.50 to 6.61 (object 61).

Family Felidae

Puma concolor.- It is commonly known as puma or mountain lion, is other one of the large Mexican carnivorous and its widely distributed throughout the American continent. Of great esteem since prehispanic times, its osseous remains have been found in archaeological contexts at different sites. At Cantona, these remains were recovered in offerings, ball games and burials. In the Great Temple of Tenochtitlan, complete skeletons of this feline, its fur [4] and bone made awls can be found.

The identified pieces of this specie were six awls inside two units. In the first unit one right metatarsus II, one right metatarsus III, one right metatarsus IV and one left metatarsus III with its first phalanx were identified from the 201 Unit, East Square (extension), layer II (75-79 elements). In the second unit, a left radius and one right metatarsus III were identified from the 9 Unit, Central Square, well 1, burial 16 (element 82).

TYPOLOGICAL ANALYSIS

The worked objects were classified according to the morphological and functional proposal from which self sacrifice awls are characterized, which according to Reyes, Pérez [5] and Heyden [6] have to present the following attributes (figure 2):
- Active borders with cutting morphology which angles oscillate between five and ten degrees.
- Raw materials that belong to large carnivores or birds of prey.
- The objects are found in public spaces, like squares, ball game structures, and burials.
- The length between the active border and the opposite border oscillates between 60.7 to 140 mm.
- The worked traces that this objects present is a polish on the active zone.

Thanks to the XVI century historical sources [7-8] it is known that for the Mexicas the self sacrifice artifacts were reserved only for the ruling classes and were fabricated from jaguar and eagle bones. These awls were used for piercing body parts, specially ears, lips, tongue, calves, the arms and the genital organ. Afterwards, the subject extracted blood from the pierced part in a way of penitence. It should be noted that we are taking these data to infer how the awls were used by the people who lived in Cantona.

TECHNOLOGICAL ANALYSIS

The manufacturing techniques of the awls were analyzed under the methodology developed under the "Manufacturing techniques of the shell objects from pre-Hispanic Mexico" project and proposed by Velázquez [9]. In this project the objects are replicated experimentally, using techniques and tools that are assumed to have been used on prehispanic times. The purpose is to employ the materials and tools which, through different sources (archaeological findings, historical sources and studies by other researches), are assumed were used in the past.

In order to avoid the speculative level and to propose more accurately the tools and techniques employed, the manufacturing traces obtained in the experiments are characterized and compared with the archaeological ones through three levels of analysis: macroscopic (plain sight), optic microscopy at low amplifications (10X, 30X and 63X) and scanning electron microscopy (SEM) (100X, 300X, 600X and 1000X). The latter is the technique that has achieved the best results, for it is suitable in the studies of the surface characteristics of the materials.

Figure 2. The image of a character using a awl of self sacrifice (Florentine Codex by Franciscan friar Bernardino de Sahagún).

For the technological analysis of the awls collection from Cantona, 15 replicas with different modifications, such as surfaces and cuts, were obtained. These polymers were coated with gold ions and observed in high vacuum mode (HV), with a beam energy of 20 kV, an spotsize of 42, a work distance of 10 mm, with a secondary electrons signal (SEI), and four amplifications (100X, 300X, 600X and 1000X) of each were obtained. Later, the micrographs were compared with experimental traces made in modern materials and with the tools indicated by the analyzed traces in the micrographs currently counted in the database of the "Manufacturing Techniques of bone objects from pre-Hispanic Mexico Project", obtaining the following results for:

Surfaces: 11 awls surfaces polymers were analyzed:
- Puma (*Puma concolor*): right radius, right II metatarsus, right III metatarsus and right IV metatarsus IV.
- Wolf (*Canis lupus bailey*): right III metatarsus.

Cuts: Three polymers were analyzed for:
- Puma (*Puma concolor*): right radius and right IV metatarsus.
- Wolf (*Canis lupus baileyi*): right III metatarsus.

Surfaces

The results of observation under SEM indicate that in the case of the surface abrading, it was possible to appreciate 21 to 33 μm bands, which get stacked and most of them were parallel to each other; these traces matches with the experimental abrading with rhyolite (figure 3 and 4).

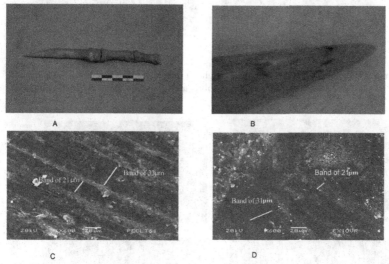

Figure 3. The image of a awl elaborate of a right III metatarsus of a wolf (A). An amplified image of awl 40x (B). It was abraded with rhyolite: archaeological (C) and experimental (D).

Cuts

For the SEM studies, these modifications a succession of very fine lines of 0.7 μm of width could be seen, these traces match the ones produced by cutting or making incisions with sharp obsidian tools (figure 5).

257

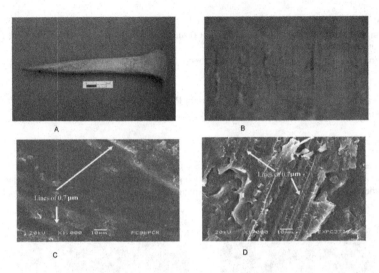

Figure 4. The image of a awl elaborate of a right III metatarsus of a puma (A). An amplified image of awl 40x (B). It was abraded with rhyolite: archaeological (C) and experimental (D).

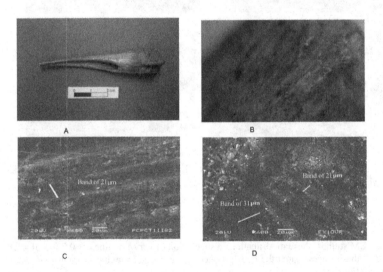

Figure 5. The image of a awl elaborate of a right radius of a puma (A). An amplified image of awl 40x (B). It was cut with obsidian: archaeological (C) and experimental (D).

RESULTS AND DISCUSSION

The obtained results show the way that the awls were elaborated. In the first stage, they were cut with obsidian tools, and later they were abraded and regularized with rhyolite. In the case of the polymers obtained near the cutting points of the awls, the traces of obsidian and rhyolite were flattened, possibly due to the worked surface.

With the information from the context, raw material, manufacturing and wear traces, we could maintain that the studied objects are sacrifice punches.

In Cantona, the place where more puma punches (made over the back leg on four different metatarsus) come from is at the East Square (Unit 201) layer II. Perhaps these objects were left there after the rite. Other place is in the Central Square (Unit 9) burial 16, where two awls of the same carnivore (the right radius and the III metatarsus) were found; due to this evidence we can assume that the artifacts were deposited there as part of the individual offering. As for the case of the Burial 23 from the Structure 1 at the Ball Game Complex 5, one wolf awl was located; it is a right toe from the back leg that is topped with a cutting end in the opposite part of the claw.

With all these exposed data, we can corroborate that the puma (*Puma concolor)* and the grey wolf (*Canis lupus baileyi),* two large carnivores, were of great appreciation and esteem for the high rank groups at Cantona, where its bones were transformed into rituals objects.

ACKNOWLEDGEMENTS

We thank the Cantona Archaeological Project and professor Ángel García Cook, for providing the material for the present work; Engineer Gerardo Villa for his collaboration with the SEM micrographs; Edsel R. Robles for his support with the experiments; Dr. Adrián Velázquez for his advisory assistance and comments for this investigation, and the "Manufacturing techniques of the shell objects from pre-Hispanic Mexico" project.

REFERENCES

1. K. Vakiimes Serret, Á. García Cook, Cantona y sus ofrendas de concha, in: *Ecos del pasado: Los moluscos arqueológicos de México.* México, L. Suárez Diez & A. Velázquez Castro (coords.) Instituto Nacional de Antropología e Historia, INAH, Colección científica 572, Mexico, 2010, 219- 237.
2. J. Talavera, J.Martín, E. García, *Modificaciones culturales en los restos óseos de Cantona, Puebla. Un análisis bioarqueológico.* Serie Arqueología, INAH, México, 2001.
3. E. D. Wilson, D. Reeder, *Mammals species of the world. A taxonomic and geographic referente,* Smithsonian Institution Press, Washington,1993.
4. N. Valentín Maldonado, B. Zúñiga Arellano, La fauna de la ofrenda 102 del Templo Mayor de Tenochtitlan, in Arqueología e historia del Centro de México, Homenaje a Eduardo Matos Moctezuma, L. López Luján, D. Carrasco y L. Cué (Coords.), Instituto Nacional de Antropología e Historia, INAH, México, 2006.
5. I. Reyes Carlo, G. Pérez Roldán, Los punzones experimentales, un caso de estudio, in *Actualidades Arqueológicas. Pasado en presente. Arqueología experimental,* Instituto de Investigaciones Antropológicas, Universidad Nacional Autónoma de México, Mexico, 2005 (electronic publication in CD).

6. D. Heyden, Autosacrificio prehispánico con púas y punzones, *Boletín INAH*, Época II (1972) 27-30.
7. F. D. Durán, *Historia de las Indias de la Nueva España e islas de tierras firmes*, Ed. Porrúa, México, 1976.
8. F. B. de Sahagún, *Historia General de las cosas de la Nueva España* (6th. Edition by Á. Ma. Garibay), Editorial Porrúa, México, 1985.
9. A. Velázquez Castro, *La producción especializada de los objetos de concha del Templo Mayor de Tenochtitlan.* Instituto Nacional de Antropología e Historia, INAH, Colección científica 519, México, 2007.

Methodologies and Instrumentation

Mater. Res. Soc. Symp. Proc. Vol. 1374 © 2012 Materials Research Society
DOI: 10.1557/opl.2012.1395

Silver Nanoparticles for SERS Identification of Dyes

Edgar Casanova-González[1], Angélica García-Bucio[1], José Luis Ruvalcaba-Sil[1], Víctor Santos-Vasquez[2], Baldomero Esquivel[3], María Lorena Roldán[4], Concepción Domingo[4]
[1]Departamento de Física Experimental, Instituto de Física, Universidad Nacional Autónoma de México, Circuito de la Investigación Científica s/n, Ciudad Universitaria, Mexico DF 04510, Mexico. e-mail: sil@fisica.unam.mx
[2]Coordinación Nacional de Conservación del Patrimonio Cultural, INAH, Mexico.
[3]Instituto de Química, Universidad Nacional Autónoma de México, Mexico.
[4]Instituto de Estructura de la Materia, CSIC, Serrano 121, 28006-Madrid, Spain.

ABSTRACT

Coinage metals nanoparticles have been widely used in last decade for enhancing the Raman signal of a variety of compounds. Several preparation methods have been proposed, including chemical reduction of gold or silver salts with sodium citrate, hydroxylamine or sodium borohydride, microwave-assisted reduction with glucose, Tollens mirror, electrodeposition, vacuum evaporation and pulsed-laser deposition.

In this work, gold and silver nanoparticles were prepared by chemical reduction with sodium citrate and hydroxilmanine, characterized by UV-Vis spectroscopy and High Resolution Transmission Electronic Microscopy and tested as SERS substrate. Carminic acid, cochineal, axiote, muitle and zacatlaxcalli SERS spectra were recorded at different pH. Natural dyes samples were prepared by extraction from its natural sources, following traditional recipes. Although differs for each dye, best results were achieved by performing SERS experiments at pH neutral to basic.

INTRODUCTION

Since its discovery in 1974 [1], Surface Enhanced Raman Spectroscopy (SERS) had been applied in several fields and the number of papers published each year on the subject had grown steadily to add more than a thousand in 2008 [2]. In the field of art, history and archaeology, SERS has become an alternative to classical techniques for the identification of organic materials, mainly dyes, in objects of artistic or historical interest [3-6].

Non-destructive or micro-destructive techniques are needed for the study of such objects and several techniques have proven to be useful in the identification of inorganic pigments, among them X-ray Diffraction (XRD), X-ray Fluorescence (XRF), Particle Induced X-ray Emission (PIXE) and Transmission Electron Microscopy (TEM). For organic materials, on the other hand, their non-invasive characterization is difficult, since high-performance liquid chromatography (HPLC) and gas chromatography (GC) [7-9], the most common analytical techniques in organic chemistry, generally require a relatively large sample.

Raman spectroscopy has been successfully applied to the study of artistic objects [10-12], but for organic dyes the technique shows some limitations regarding signal intensity and interference from the fluorescence [13]. Some alternatives have been proposed to overcome these limitations, mainly Subtracted Shifted Raman Spectroscopy [14, 15], and Surface Enhanced Raman Spectroscopy (SERS). In SERS, the Raman signal is enhanced and the fluorescence is quenched when the analyte is close to or absorbed on metal nanostructures [14].

The first paper dealing with the identification of organic dyes by SERS was published in 1987 [16]. Since then, much effort had been devoted to the study of pure dyes and artwork samples [17-25]. On the contrary, from the more than twenty known Mexican dyes, only a handful have been studied by Raman or SERS, the list including only indigo [26], cochineal [27], brazilwood [28] and achiote [29]. The most known are indigo (present in the famous Maya blue) and carminic acid, the molecule responsible for the red color of cochineal red. The rest of the dyes are mainly reds and yellows, with blues and blacks also included.

As a starting point, we decided to prepare and characterize silver and gold colloids, and then test them as SERS substrates using pure carminic acid, to choose the best suitable colloid. The selected colloid was then used to acquire the SERS spectra of dye extracts of cochineal, achiote, muitle and zacatlaxcalli, looking for the best pH conditions.

EXPERIMENT

Sample preparation

Fresh samples of cochineal, achiote, muitle and zacatlaxcalli were collected or bought in local markets. After cleaning, crushing (only for cochineal) and drying, they were macerated with water and left alone for three days. The cochineal extract was filtered and then freeze-dried. The remaining extracts were also freeze-dried.

Synthesis of silver and gold colloids

Citrate-reduced silver and gold colloids: A solution of 10^{-3} molL^{-1} AgNO$_3$ (or 2.24 x 10^{-3} molL^{-1} HAuCl$_4$) was prepared and 50 mL of this solution was heated to reflux under stirring. Then, 1 mL of sodium citrate 1 % solution was added and the mixture was kept under reflux for 60 min (5 min for gold). The reaction mixture was left to cool to room temperature and left to rest for a day.

Hydroxilamine-reduced silver colloids: In a 250 mL Erlenmeyer were poured 90 mL of a 1.66 x 10^{-3} molL^{-1} solution of NH$_2$OH.HCl and stirred to 300 rpm before adding 300 µL of NaOH. Under stirring, 10 mL of 10^{-2} molL^{-1} AgNO$_3$ were added dropwise for one minute and the mixture were stirred at low speed for ten more minutes and left to rest for a day.

All three colloids were characterized by UV-Vis Spectroscopy and Transmission Electron Microscopy. UV-Vis spectra were acquired on a Shimadzu UV 3600 spectrometer. TEM images were acquired on a JEOL 2010 FasTem, FEG type, electronic microscope, with an acceleration voltage of 200 kV. Both operation modes TEM and STEM were used for clear field and Z-contrast (HAADF) images, respectively. Images were analyzed using the ImageJ package.

Surface Enhanced Raman Spectroscopy SERS

SERS spectra were recorded on a Delta Nu Inspector Raman, equipped with a 785 nm laser and a maximum power of 100 mW. Laser level was set to high (100 mW), the range was 200 to 2000 cm^{-1}, the integration time was 2 s and 15 spectra were averaged on each experiment. When working with aqueous solutions, 900 µL of silver colloid were mixed with 40 µL of 0.5 molL^{-1} KNO$_3$ and then 100 µL of dye sample (10 mgmL^{-1}) were added. The pH of the mixture was adjusted with 0.1 molL^{-1} NaOH and 0.1 molL^{-1} HCl.

DISCUSSION

Silver colloids

The first reported SERS substrate was a silver electrode, used by Fleischmann in 1974 to study the adsorption of pyridine molecules at electrodes by acquiring their Raman spectra [1]. Since then, SERS substrates had been prepared in multiple ways, commonly by reduction of silver or gold salts with sodium citrate [30-33], hydroxylamine [34] or sodium borohydride [32, 35]. Tollens mirror [34], vacuum deposition [34], synthesis of gold nanorods [36], microwave-assisted reduction [27], and pulsed-laser deposition [37] are some of the methods also employed to prepare SERS-active nanostructures.

The Lee-Meisel colloid has several advantages, namely the simplicity of its synthesis and its good behavior as SERS substrate [5], but it is difficult to prepare colloids batches with the same SERS activity maintaining unchanged the reaction conditions [19]. Both simplicity and SERS activity are also the advantages of the hydroxylamine-reduced silver colloid and the citrate-reduced gold colloid. All three colloids were synthesized and characterized by Transmission Electron Microscopy and UV-Vis spectrometry (Figure 1 and 2).

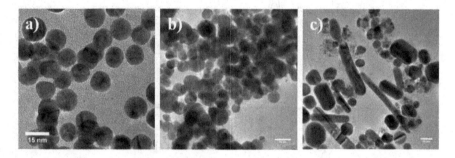

Figure 1. TEM images of metal colloids. a) citrate-reduced silver colloid; b) hydroxylamine-reduced silver colloid; c): citrate-reduced gold colloid

From the UV-Vis absorption spectra, it was apparent that all three colloids were suitable as SERS substrate, but the TEM images showed a greater size dispersion hydroxylamine-reduced silver colloid and the citrate-reduced gold colloid. The gold colloid showed also a great diversity of shapes, as opposed to the two silver colloids, were almost all of the particles had spherical shapes.

When their SERS activity was tested, using a 10^{-2} molL^{-1} solution of carminic acid and a 785 nm excitation laser, only the citrate-reduced silver colloid gave useful SERS spectra and was chosen for the remaining test with the dye extracts. This colloid has been stable for over a year, and shows a UV-Vis absorption spectrum centered in 409 nm, with a FWHM of 137 (see Figure 2). The high resolution transmission electronic microscopy revealed a very narrow size distribution, centered in 14 nm of diameter. Although sodium chloride and potassium nitrate were tested as aggregation agents and gave similar results (Figure 2), potassium nitrate was chosen based on previous attempts to achieve SERS with added NaCl, which proved unsuccessful.

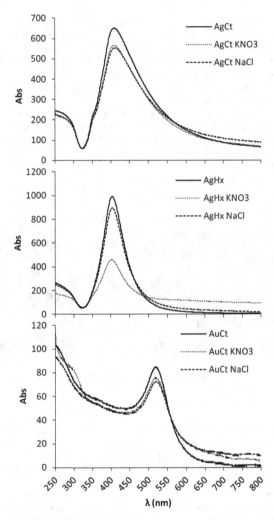

Figure 2. UV-Vis absorbtion spectra of silver and gold colloids. AgCt: citrate-reduced silver colloid; AgHx: hydroxilamine-reduced silver colloid; AuCt: citrate-reduced gold colloid.

SERS of dyes in aqueous solution

The SERS spectra of carminic acid were recorded at pH levels ranging from 2 to 13 (Figure 3). At acid pH levels, no specific bands of the molecule were observed, instead, a broad band is present, centered in $900 - 1000$ cm^{-1} and probably due to the fluorescence of the acid. This fluorescence might be related to a poor absorption of the molecule on the silver, perhaps because of the interference of the chloride ions added to the mixture when using HCl for pH adjustment. At pH 6, a drastic change is observed and the most intense spectrum is observed at pH 7. Subsequent increases on pH lead to less intense spectra and the good results obtained at neutral pH conditions should be due to the molecule structure, since carminic acid is an athraquinone with several hydroxyl substituents, two ketones and a carboxyl groups and a glucose residue. This molecule acquires a more negative overall charge as the pH gets more alkaline and thus its absorption on the surface of the silver nanoparticle becomes more and more difficult, since this surface presents a high negative zeta potential [28]. The SERS spectra showed a very strong band at 1290 cm^{-1}, related to the $v(CC)$, $\delta(C_5OH)$, $\delta(C_8OH)$ and $\delta(C_3OH)$ vibrations, as well as other bands at 1077 cm^{-1} and 462 cm^{-1}, related vibrations of the glucose residue and the carbon chain, respectively [28].

After these first results, all the remaining SERS spectra were recorded at neutral or alkaline pH. For the cochineal extract, best results were achieved at pH 7 and the signal intensity decreased when higher pH levels were reached, as seen on Figure 4. The carminic acid bands previously observed were also present in the cochineal spectra.

For the achiote extract, a higher pH was needed in order to obtain an intense and well defined. At pH levels from 6 to 8, although the reported [30] bands of bixin can be observed, they are broad and with low intensity (Figure 5). At pH 11 and higher, however, very intense and well defined bands can be observed at 1007, 1153, 1183 and 1528 cm^{-1}. This need for strong alkaline conditions might be explained if we take a closer look at the SERS spectra, particularly the bands at 1153 and 1183 cm^{-1}. The peak at 1183 cm^{-1} is related to the vibrations of the C-H bond and its intensity diminishes as the peak at 1153 cm^{-1} – related to the vibrations of the C-C bond – gets more intense, which might indicate the occurrence of a nucleophilic substitution under strong basic conditions. Such reaction could be an aldol condensation and if it indeed occurs, the coupling of two bixin molecules increases the number of available donor groups and facilitates the absoption of the molecule on the surface of the silver nanoparticles.

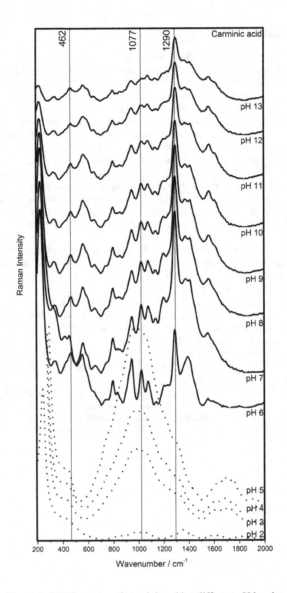

Figure 3. SERS spectra of carminic acid at different pH levels.

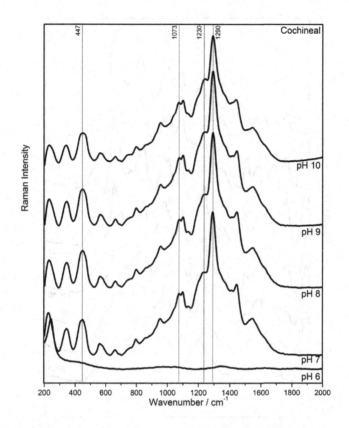

Figure 4. SERS spectra of cochineal extract at different pH levels.

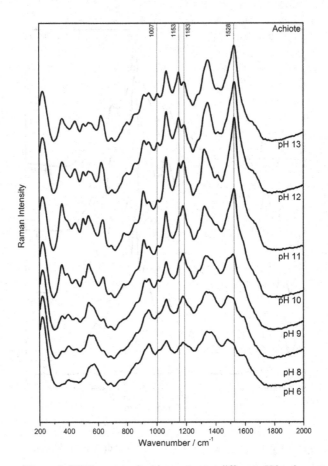

Figure 5. SERS spectra of achiote extract at different pH levels.

The structure of the main colorant in muitle an zacatlaxcalli have not yet been fully determined, muitle have been said to contain indigo as its main dye [38] but the presence of the anthocyanin kaempferitrin had been found on its leaves [39] and this is coherent with the color obtained when dyeing with this plant. Recent studies by Esquivel *et. al.* (unpublished) points to the presence of the flavonoid quercetin in zacatlaxcalli extracts. For both muitle and zacatlaxcalli, the presence of several bands in the $1100 - 1600$ cm^{-1} region and some very strong bands near 500 cm^{-1} agrees with these findings and suggest a flavonoid like structure [26, 40]. Best spectra were acquired at neutral pH, as in the case of carminic acid and it is probably linked to the presence of available electron pairs (hydroxyl, ketone or carboxyl groups) that allows a good absorption of the analyte and therefore intense SERS spectra (Figures 6 and 7).

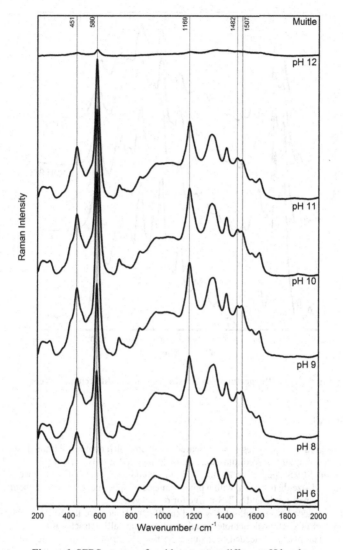

Figure 6. SERS spectra of muitle extract at different pH levels.

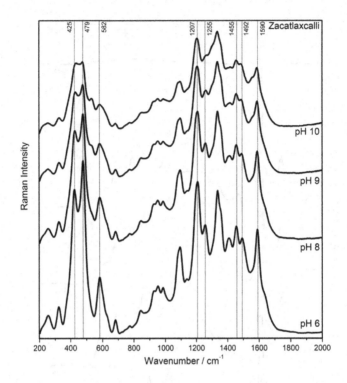

Figure 7. SERS spectra of zacatlaxcalli extract at different pH levels.

CONCLUSIONS

Silver and gold nanoparticles were prepared by chemical reduction. Although the UV-Vis spectra suggested that all three colloids were suitable for SERS, only the citrate-reduced colloid resulted in good SERS spectra under the 785 nm excitation. This colloid showed the most uniform shape and size distribution, with a predominance of spherical particles of about 14 nm in diameter and remained stable and SERS-active for over a year.

The citrate-reduced silver colloid was used to effectively acquire the SERS spectra of pure carminic acid and of cochineal, achiote, muitle and zacatlaxcalli extracts, at neutral to alkaline pH conditions. The pH level needed to acquire the best spectrum can be related to the known chemical structure of carminic acid and bixin, for cochineal and achiote. In the case of muitle and zacatlaxcalli, although their chemical structure has not yet been fully elucidated, the SERS spectra showed features that have been previously observed on flavonoids.

ACKNOWLEDGMENTS

This research has been supported by CONACyT MOVIL II grant 131944 and PAPIIT UNAM project IN403210 ANDREAH. E. Casanova received support from CONACyT fellowship program for his PhD research in the frame of the Posgrado en Ciencia e Ingenieria de Materiales, Universidad Nacional Autonoma de México (UNAM).

REFERENCES

1. M. Fleishmann, P.J. Hendra, A.J. McQuillan, *Chem. Phys. Lett.* 26 (1974) 163.
2. D. Graham, R. Goodacre, *Chem. Soc. Rev.* 37 (2008) 883
3. K. Chen, M. Leona, K.C. Vo-DinhF. Yan, M.B. Wabuleye, T. Vo-Dinh, *J. Raman Spectrosc.* 37 (2006) 520.
4. M. Leona, J. Stenger, E. Ferloni, *J. Raman Spectrosc.* 37 (2006) 981.
5. A.V. Whitney, R.P. Van Duyne, F. Casadio, *Proc. of SPIE* 5993 (2005) 117.
6. A.V. Whitney, R.P. Van Duyne, F. Casadio, *J. Raman Spectrosc.* 37 (2006) 993.
7. E. Rosenberg, *Anal. Bioanal. Chem.*, 391 (2008) 33.
8. E. de Rijke, P. Out, W.M.A. Niessen, F. Ariese, C. Gooijer, U.A. Th. Brinkman, *J. Chromatogr.* A 1112 (2006) 31.
9. X. Zhang, R. Boytner, J.L. Cabrera, R. Laursen, *Anal. Chem.* 79 (2007) 1575.
10. S.P. Best, R.J.H. Clark, R. Withnall, *Endeavour* 16 (1992) 66.
11. R.J.H. Clark, *J. of Mol. Structure* 347 (1995) 417.
12. A. Adriaens, *Spectrochim. Acta*, Part B 60 (2005) 1503.
13. M. Moskovits, *J. Raman Spectrosc.* 36 (2005) 485.
14. F. Rosi, M. Paolantoni, C. Clementi, B. Doherty, C. Miliani, B.G. Brunetti, A. Sgamellotti, *J. Raman Spectrosc.* 41 (2010) 452.
15. S.E.J. Bell, Case Study: Chinese Scrolls and other Fluorescent Samples, in *Raman Spectroscopy in Archaeolgy and Art History*, chapter 18, H.G.M Edwards & J.M. Chalmers (eds.), The Royal Society of Chemistry, Cambridge, UK, 2005.
16. B. Guineau, V. Guichard, *ICOM Committee for Conservation: 8th Triennial Meeting, vol 2.* The Getty Conservation Institute, Los Angeles, 1987, 6.
17. B.H. Berrie, *PNAS* 106:36, (2009) 15095.
18. M. Leona, *PNAS* 106:35, (2009) 14757.
19. N. Peica, W. Kiefer, *J. Raman Spectrosc.* 39 (2008) 47.
20. C.L. Brousseau, A. Gambardella, F. Casadio, C.M. Grzywacz, J. Wouters, R.P. Van Duyne, *Anal. Chem.* 81 (2009) 3056.
21. M.V. Cañamares, M. Leona, M. Bouchard, C.M. Grzywacz, J. Wouters, K. Trentelman, *J. Raman Spectrosc.* 41 (2010) 391.
22. C.L. Brousseau, K.S. Rayner, F. Casadio, C.M. Grzywacz, R.P. Van Duyne, *Anal. Chem.* 81 (2009) 7443.
23. K.L. Wustholz, C.L. Brousseau, F. Casadio, R.P. Van Duyne, *Phys. Chem. Chem. Phys.* 11 (2009) 7350.
24. S. Bruni, V. Guglielmi, F. Pozzi, *J. Raman Spectrosc.* 41 (2010) 175.
25. Z. Jurasekova, E. del Puerto, G. Bruno, J.V. García-Ramos, S. Sanchez-Cortes, C. Domingo, *J. Raman Spectrosc.* 41 (2010) 1165.
26. P. Vandenabeele, S. Bodé, A. Alonso, L. Moens, *Spectrochim. Acta*, Part A 61 (2005) 2349.

27. M.V. Cañamares, J.V. Garcia-Ramos, C. Domingo, S. Sanchez-Cortes, *Vibrat. Spectrosc.* 40 (2006) 161.
28. L.F.C. de Oliveira, H.G.M. Edwards, E.S. Velozoc, M. Nesbitt, *Vibrat. Spectrosc.* 28 (2002) 243.
29. L.F.C. de Oliveira, S.O. Dantas, E.S. Velozo, P.S. Santos, M.C.C. Ribeiro, *J. Mol. Struct.* 435 (1997) 101.
30. P.C. Lee, D. Meisel, *J. Phys. Chem.* 86 (1982) 3391.
31. R. Aroca, *Surface-Enhanced Vibrational Spectroscopy*, John Wiley & Sons, Chichester, 2006.
32. L. Rivas, S. Sánchez-Cortes, J.V. García-Ramos, G. Morcillo, *Langmuir* 17 (2001) 574.
33. Z. Wang, S. Pan, T.D. Krauss, H. Du, L.J. Rothberg, *PNAS* 100 (2003) 8638.
34. N. Leopold, B. Lendl, *J. Phys. Chem.* B. 107 (2003) 5723.
35. R.F. Aroca, R.A. Alvarez-Puebla, N. Pieczonka, S. Sanchez-Cortez, J.V. Garcia-Ramos, *Adv. Colloid Interface Sci.* 116 (2005) 45.
36. C.J. Murphy, T.K. Sau, A.M. Cole, C.J. Orendorff, J. Gao, L. Gou, S.E. Hunyadi, T. Li, *J. Phys.Chem.* B 109 (2005) 13857.
37. S. Sánchez-Cortés, L. Guerrini, J.V. García-Ramos, C. Domingo, *J. Phys.Chem.* B 111 (2007) 8149.
38. D.R.A. Watson, *The Book and Paper Group Annual*, The American Institute for Conservation, Washington DC, 2005, 137.
39. K.L Euler, M. Alam, *J. Nat. Prod.* 45 (1982) 220.
40. Z. Jurasekova, A. Torreggiani, M. Tamba, S. Sanchez-Cortes, J.V. Garcia-Ramos, *J. Mol. Struct.* 918 (2009) 129.

Mater. Res. Soc. Symp. Proc. Vol. 1374 © 2012 Materials Research Society
DOI: 10.1557/opl.2012.1396

Photoacoustic Analysis of Natural Indigo, Palygorskite and Synthetic Maya Blue

Carlos Aldebarán Rosales Córdova[1], Antonio de Ita de la Torre[1], Rosalba Castañeda Guzman[2]

[1] Área de Ciencia de Materiales, Universidad Autónoma Metropolitana UAM.
Av. San Pablo No. 180, Col. Reynosa Tamaulipas, Mexico DF C.P. 02220, Mexico.
e-mail: add@correo.azc.uam.mx
[2] Centro de Ciencias Aplicadas y Desarrollo Tecnológico CCADET, Universidad Nacional Autónoma de México UNAM, A.P. 70-186 Mexico D.F C.P. 04510, Mexico.

ABSTRACT

Maya blue is an organic / inorganic pigment which is composed principally by two elements: a white clay known as palygorskite (in the Mayan dialect Sakalum), and a blue dye called indigo, this dye is extracted from the plant *Indigofera suffruticosa*. The mixture and the warming of these two elements produce the Maya blue, which exhibits unusual features such as: resistance to the assault of the acids in warm or high temperatures, his persistent color in spite of having been exposed to different climatic conditions as a result of the passage of time and the specific chemical composition that this pigment presents.

In the present investigation natural indigo, palygorskite and synthetic maya blue were analyzed with a new implementation technique called photoacoustic analysis, which detects the structural changes that happen in the material under a controlled increase of temperature. Due to the fact that this technology detects with clarity where the structural changes happen but not that type of changes happen, it used as base termogravimetric analysis. One of the important findings, it was that in the spectrum of the synthetic maya blue were detected the structural changes of the clay and the dye, something that with other used technologies had not been achieved to observe.

INTRODUCTION

Merwin in 1931 discovers a pigment with a blue tone turquoise on having done an expedition to the temple of the "Warriors", which is located in the archaeological zone that today is known as Chichén Itzá, in the state of Yucatan [1]. In 1942 - one year later - the above mentioned pigment was baptized as Mayan Blue by Gettens [2], given the combination of the tone - color - and the representative local civilization where it was discovered.

From this year, it began the study of the Maya blue analyzing the paintings that Merwin had found. The first question that arose for the different investigators and scientists was: which are the elements that compose this characteristic pigment? It was then determined that the pigment is formed by a singular association between clays and vegetable inks, being a result of the mixture of two principally components: Paligorskite —clay— and the dye called indigo contained in the leaves of the plant *Indigofera suffruticosa*.

Palygorskite is a fibrous clay which chemical formula is [$Mgi_2S_4O_{10}(OH)$ $4 \cdot H_2O$], its average dimensions are between 0.1 and 2 mm, with a width between 100 to 300 Å, it has a structure anisotropic with porous channels and dimensions of 4.1 Å of width x 10.5 Å of high and 16.2 Å of depth. On the other hand, the natural indigo has a blue very intense color, is practically insoluble in any solvent, his structure is monoclinic in crystalline phase and his molecule presents the following dimensions: 14 Å of length, 5.3 Å of width and 3.4 Å of thickness [3].

Maya blue has drawn attention from different disciplines of science such as archeology, anthropology, engineering, and many others, due to the fact that this pigment presents slightly common characteristics, which are: Its high resistance to the assault of the acids concentrated - sulphuric, nitric and hydrochloric acid [2, 4]. The unusual chemical stability - which has created a scientific debate, with theories that suggest that the indigo seals the channels of the palygorskite [5, 6] or penetrates inside them [7-9]. Its composition: the strange association that form the mixture of an organic material - indigo - with an inorganic material - palygorskite. The persistence of the color in spite of being exposed to different climatic conditions as a result of the passage of time. All these present characteristics in the Maya blue are unusual, since other blue pigments do not present them or of another color used by the different civilizations of the world.

From his discovery many methods of characterization have been applied to Maya blue by the intenseness of revealing the nature of this pigment and of gaining information about the form in which the thermal stability as well as the manufacturing process takes place. The techniques most mentioned by the investigators for the Maya blue studies include: Particle Induced X-ray Emission (PIXE) [10]; X-ray Absortion Spectroscopy (XAS) [11]; High Resolution Transmission Electron Microscopy (HRETM) [12]; X-ray Fluorescence (XRF) [13]; Raman spectroscopy [13]; Differential Thermal Analyses (DTA, DSC, TGA) and X-ray Diffraction (XRD).

The use of the techniques of characterization mentioned above have served to answer many questions that raised about the pigment, but still unanswered questions remain inconclusive, one of them is: which is the reason why indigo is not observed in the spectrum of Maya blue? For this reason, in the present investigation there was used a new technique of characterization called Photoacoustic analysis. This method consists basically of the detection of structural changes (with high sensibility) by means of a pulsed laser which interact with the sample generating photo-acoustic signs.

EXPERIMENTAL

Photoacustic analysis:

Photoacoustic analysis lies on generating acoustic waves in the material from the excitation with light of pulsed laser —pulse of 7 ns— and piezoelectric sensors to detect the acoustic signals. This methodology has shown a high sensibility and a high resolution with a high signal-noise ratio. The photo-acoustic signals of a sample depend in the method of direct coupling [17-19] on the thermal expansion ΔV_{th} of the radiated volume V_0, which for the isotropic case is:

$$\Delta Vth = \frac{\beta}{\rho Cp} H$$

where β is the volumetric expansion coefficient, C_p is the specific heat at constant pressure, ρ is the density and H is the heat deposited in the volume V_0. This local thermal expansion generates a pressure wave which travels at the speed of sound through the material. Each laser pulse generates a photoacoustic signal representative of the material in the thermodynamic state prevails, and will not change until an external parameter, in this case the temperature modified the structure of the material.

Figure 1 shows the scheme that shows the procedure that is used to realize the photoacoustic measurements.

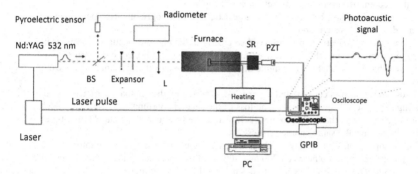

Figure 1. Scheme of the experimental design used for the photoacoustic measurements. BS: Divisor of bundle, L: lens, SR: System of refrigeration.

Thermogravimetric analysis:

In dinamic termogravimetry, the sample is heated so that the oven temperature is varying linearly. A material is thermally stable if it does not change when the temperature increase or decrease, a measure of this stability is the variation of the mass. In thermal differential analysis the results that are obtained will depend on two types of several factors: the set of instruments and that of the sample itself. This technique consists of observing basically the loss of mass that happens in the material during the warming with a program of the variation of the temperature.

Materials and samples

Palygorskite clay was brought from the village of Tikul, located in the State of Yucatan. It is one of the most important and known sources. On the other hand Indigo was acquired in the state of Oaxaca from local producers.

Synthetic Maya blue was prepared using the paligorskite and the indigo in order to obtain two samples: A first preparation with a ratio of 5 g clay for 0.05 g dye and a second one with a ratio of 5 g clay for 0.5 g dye. To prepare the synthetic Maya blue, the clay and the indigo were ground in a mortar of porcelain, later a mechanical mixture was done between both elements and the mixture was warmed during 30 minutes at 200 °C.

RESULTS AND DISCUSSION

Figure 2 shows the result of thermogravimetric and photoacoustic analysis of the clay, in which four principal changes are observed. First: the loss of the zeolitic water that appears below the 100 °C, second: the loss of the first coordinated water between 180 and 250 °C, third: the loss of the second coordinated water between 450 and 550 ° C, and finally, the decomposition of the clay between 700 and 900 ° C.

After the comparison between the photoacustic and thermogravimetric spectra, three important differences were found:

a) Loss of zeolitic water is not observed so clearly in the photo-acoustic spectrum as in the thermogravimetric one, this is because the zeolitic water is not linked to the structure of the clay, which prevents that the technique detect it clearly —despite this, small peaks were observed.

b) Loss of the second coordinated water is not detected by the photoacoustic technique, generating new research questions as: The loss of the second coordinated water really does not involve a structural change? If so, how is linked to the structure of the clay? Other types of palygorskite have the same characteristics?

c) Decomposition of the clay is observed as a series of peaks in the photoacoustic spectrum in the temperature between 700 and 900 ° C, this structural change is not detected by thermogravimetric analysis because it does not involve a loss of mass.

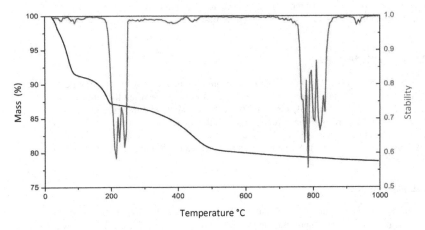

Figure 2. Photoacustic and thermogravimetric analyses of the palygorskite

Figure 3 shows the result of thermogravimetric and photoacoustic analysis of the dye. Three main changes were found: the loss of adsorbed water that occurs below 110 ° C, sublimation which is between 200 and 400 ° C, and the decomposition which start at 520 °C. Comparing the two spectra, there are two significant differences: a) loss of adsorbed water is not clearly detected by photoacoustic technique, b) the decomposition of dye was observed with the photoacoustic technique as a series of peaks starting from 520 ° C , while the thermogravimetric technique shows it gradually.

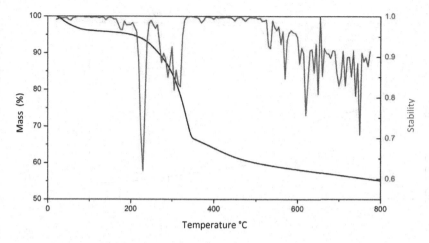

Figure 3. Photoacustic and thermogravimetric analysis of indigo sample.

Figure 4 shows the result of thermogravimetric and photoacoustic analysis of Maya blue (5:0.05 ratio). There are four major changes, first: the loss of zeolitic water that occurs below 100 °C, second: the loss of the first coordinated water which is between 190 and 300 ° C, third: the loss of the second coordinated water between 380 and 500 ° C, fourth: the decomposition of the pigment between 750 and 900 ° C.

The spectrum of Maya blue is very similar to the clay, as reported by different authors, the most significant difference is the well defined peak that is observed at 400 °C in the photoacoustic spectrum, leading to generate the hypothesis that this peak is not simply due to the loss of the second coordinated water, because when the palygorskite was analyzed alone its spectrum did not show that change.

Figure 5 presents the structural changes detected by photoacoustic and thermogravimetric techniques on synthetic Maya blue (5:0.5 ratio). It was found that there is no change compared with the specter of the other synthetic Maya blue (5:0.05 ratio) in the first two changes detected, which are due to the loss of zeolitic water and first coordinated water, but in the third structural change detected by the photoacoustic technique there are two series of peaks between 380 and 600 ° C.

Such changes were observed as one peak in the spectrum of Maya blue indigo with less amount of indigo, it is believed that the first series of peaks are due to the sublimation of indigo, while the second series of peaks are due to the loss of the second coordinated water coordinated, showing the ability of the technique (due to its high sensitivity) to detect structural changes of indigo in the spectrum of Maya blue, situation that different techniques of characterization have not found [3, 9-10]. Despite these results, different questions have emerged, such as: the first series of peaks are really due to the sublimation of indigo and the second to the loss of the second first coordinated water? How much amount of indigo is necessary in the production of Maya blue to start to observe the sublimation? These are the questions that will try to answer in future investigations.

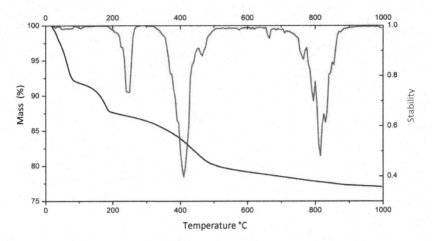

Figure 4. Photoacustic and thermogravimetric analysis of Maya blue (5:0.05 ratio)

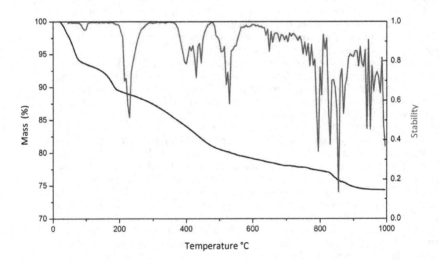

Figure 5. Photoacustic and thermogravimetric analysis of Maya blue (5:0.5 ratio).

CONCLUSIONS

The loss of the second coordinated water is not clearly detected by the photoacoustic technique.

Independently of the quantity of indigo that is used to realize the Maya blue, below 300 °C the structural changes of the clay are only observed.

In the spectrum of Maya blue with more amount of indigo, the photoacoustic technique detects the dye sublimation and the loss of the second first coordinated water of the clay, that is, this method is able to reveal structural changes of the dye in the spectrum of the pigment.

Decomposition of the pigment is detected with much greater intensity in Maya blue (5:0.5 ratio) due to the photoacoustic method is able to observe together the decomposition of indigo and palygorskite.

REFERENCES

1. H. E. Merwin, in: *The Temple of the warriors at Chichen-Itza, Yucatan*, E. H. Morris, J. Charlot, A. A. Morris (eds.), Carnegie Institution,Washington D.C., 1931.
2. R.J. Gettens, *American Antiquity* 27 (1962) 557-564.
3. L.A. Polette, F.S. Manciau, B. Torres, M.A. Jr,R.R. Chianelli, *Journal of Inorganic Biochemestry* 101 (2007) 1958-1973.
4. C.R. Valerio, *De Bonampak al Templo Mayor. EL Azul Maya en Mesoamérica*, Colección América Nuestra, Siglo XXI ed., 1993.
5. B. Hubbard, W. Kuang, A. Moser, G. A. Facey, C. Detellier, *Clays and Clay Minerals* 51 (2003) 318-326.
6. H. Van Olphen, *Science* 154 (1966) 645-646.
7. G. Chiari, R. Giustetto, G. Ricchiardi, *European Journal of Mineralogy* 15 (2003) 21-33.
8. R. Kebler, R. Masschelein-Kleiner, J. Thissen, *Studies in Conservation*, 12 (1967) 81-101.
9. E. Fois, A. Gamba, A. Tilocca, *Microporous and Mesoporous Materials* 57 (2003) 263-272.
10. M. S. del Río, P. Martinetto, C. Solís, C. R. Valerio, *Nuclear Instruments and Methods in Physics Research* B 249, (2006) 628-632.
11. L. A. Polette, G. Meitzner, M. J. Yacaman, R. R. Chianelli, *Microchemical Journal 71* (2002) 607.
12. M. S. del Río, P. Martinetto, A. Somogyi, C. Reyes-Valerio, E. Dooryhée, N. Peltier, L. Alianelli, B. Moignard, L. Pichon, T. Calligaro, J.C. Dran, *Spectrochimica Acta Part B: atomic spectroscopy* 59 (2004) 1619–1625.
13. P. Vandenabeele, S. Bodé, A. Alonso, L. Moens, *Spectrochimica Acta Part A: Molecular and Biomolecular Spectroscopy* 61 (2005) 2357-2363.

Mater. Res. Soc. Symp. Proc. Vol. 1374 © 2012 Materials Research Society
DOI: 10.1557/opl.2012.1397

Site-Specific Analysis of Deformation Patterns on Archaeological Heritage by Satellite Radar Interferometry

Deodato Tapete[1] and Francesca Cigna[2]
[1]Department of Earth Sciences, University of Firenze, Via G. La Pira 4, I-50121, Firenze, Italy.
 e-mail: deodato.tapete@gmail.com
[2]British Geological Survey, Nicker Hill, Keyworth, NG12 5GG Nottingham, UK

ABSTRACT

Exploitation of satellite radar interferometry on huge cultural heritage sites can facilitate the recognition of spatially distributed deformation patterns, whose morphology, jointly with the analysis of displacement time series, could clarify the nature of ongoing deterioration phenomena threatening the conservation of exposed archaeological heritage. Radar-interpretation is used on selected case studies located in Southern Italy to demonstrate the feasibility of Persistent Scatterers (PS) analyses for site-specific detection of superficial deformation, correlated to natural and/or human-induced instability processes. Evidence of subsidence for the radar targets identified within the archaeological area of Capo Colonna, Central Calabria, confirms the susceptibility of the entire promontory to ground instability, with potential effects on the ruins. Similarly, the uplift/subsidence patterns on the monumental area of Pozzuoli, W of Naples, testify the exposure of the geologic substratum underneath the archaeological structures to the active dynamics of the Campi Flegrei volcanic complex. Finally, the satellite analysis on the Valley of the Temples in Agrigento, Sicily, exemplifies the capability to distinguish differential displacement trends and seasonal variations within single PS time series.

INTRODUCTION

On site preservation of cultural heritage can benefit from the exploitation of remote sensing techniques which can support the identification of sectors and monuments affected by conservation criticalities due to ongoing instability processes. Such preventive diagnosis, extended over huge areas of investigation, can help the planning of necessary restorations, following a priority criterion based on a deeper and updated knowledge about the condition of the built heritage. Recent experimentations [1,2] have demonstrated the promising potentials of Synthetic Aperture Radar Interferometry (InSAR) techniques for monitoring activities at different scales, i.e. 'single monument scale' and 'entire site scale', respectively, by means of ground-based and satellite configurations. The latter is highly suitable for wide-area monitoring, allowing the detection of Earth's surface changes [3], as well as natural and/or human-induced phenomena. An interesting example is offered by the displacements observed on structures and infrastructure within the urban contexts of some cities in Central Mexico and in the historic centre of the UNESCO site of Morelia, due to subsidence accelerated by groundwater overexploitation and influenced by past tectonic activity [4].

Based on site-specific analyses of satellite InSAR data available for selected archaeological sites located in Southern Italy, the paper attempts to demonstrate the usefulness of these techniques for recognition of deformation patterns which can be correlated to active deterioration processes of the structures and/or the geological substratum. The detailed analysis of deformation time series, obtained for the investigated case studies, provides a representative sample of the displacement trends which can be frequently found through InSAR data.

THEORY AND EXPERIMENTAL DETAILS

The remote sensing data tested in this work are obtained from a multi-interferogram analysis of huge archives of multi-temporal images, acquired from satellite radar sensors and processed following the principle of *Synthetic Aperture Radar* (SAR) [5]. Wide spatial coverage allows 'entire site scale' analyses to be performed on areas up to hundreds or even thousands of square kilometers, with a meter resolution on the ground.

Processing of these data with Persistent Scatterer Interferometry (PSI) leads to the detection of superficial deformation of soil and/or structures thanks to the recognition, within the processed SAR images, of reflective targets distributed on the ground, which backscatter the microwave wavelengths sent from the satellite sensors. Such reflectors act as Persistent Scatterers (PS), as they maintain their dielectric properties almost constant during the monitoring period. PS correspond to either natural or artificial elements, such as rock outcrops, boulders, buildings, infrastructure, as well as monuments, columns and archaeological ruins that stick out of the ground. Conversely, vegetated areas provide low densities or, even, absence of targets.

Displacements (Δd) occurred on a single PS in the period between two different acquisitions (Δt) are recorded at the date of satellite observation along the LOS (Line Of Sight) direction and plotted vs. time, to retrieve the graph of the displacement trend over the monitoring period (*PS deformation time series*). The measurement along the LOS, with millimeter precision on single measure, implies that each displacement value corresponds to the LOS component of the real displacement vector associated to the detected superficial deformation. The magnitude and the direction of the LOS component depend on typology and geometric configuration of the deformation affecting the observed PS, as well as on the geometry of the satellite observation, while moving along its ascending (i.e. from S to N) or descending orbit. Displacements away from the satellite result in negative values of LOS velocity (figure 1A-B); on the contrary, movements towards the satellite are expressed as positive values (figure 1C-D).

Among the deterioration processes on cultural heritage that can be detected by means of PSI, we can mention: toppling and collapses of wall masonries and columns due to instability of the geologic substratum, landslides and slope dynamics (figure 1A) or caused by natural and/or anthropogenic subsidence (figure 1B). Similarly, positive values of deformation rates may result from uplift movements due to ground expansion, clayey soil swelling, groundwater recharging, and/or toppling in the direction towards the satellite (figure 1C).

The temporal frequency with which a certain satellite observes the same area on the ground, determines the range of detectable phenomena, particularly in terms of kinetics. Shorter revisiting time generally allows more rapid processes to be monitored. The monthly sampling frequency of the radar satellites and the related periods analyzed for the case studies discussed in this paper (see table I) have configured a '*back monitoring*' activity with temporal reconstruction of past/recent deformation, facilitating up-to-date detection of patterns and trends associable to recent/ongoing local instability.

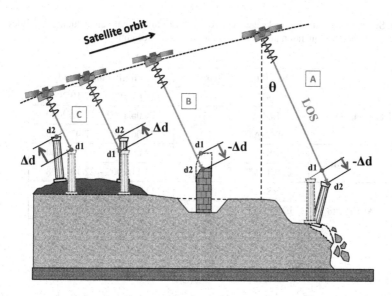

Figure 1. PSI detection of displacements (Δd) occurred on monuments between two consecutive acquisitions. Changes of PS positions (from d1 to d2) are measured along the satellite LOS (tilted from the vertical of a look angle, θ), identifying displacements away from the sensor (negative values; red arrows) due to: A) toppling; B) collapses caused by substratum instability, subsidence, earthquakes. Uplift movements due to ground expansion, clayey soil swelling and/or toppling towards the satellite (C), imply positive values of LOS deformation rates (blue arrows).

After a preliminary evaluation of the feasibility of PSI analyses on cultural heritage contexts, site-specific analyses were performed on three different archaeological sites located in Southern Italy: Capo Colonna (Central Calabria), Pozzuoli (Campania Region) and the Valley of the Temples in Agrigento (Sicily). The exploited PS data were derived from the free access database of the Extraordinary Plan of Environmental Remote Sensing (EPRS-E), created by the Italian Ministry of Environment, Territory and Sea (METS), providing satellite displacement measurements with national coverage, available for administrative and scientific purposes.

ENVISAT images, acquired with wavelength (λ) of 5.63 cm and nominal revisiting time of 35 days, were processed with the PSP-DIFSAR approach (Persistent Scatterers Pairs - DIFferential InSAR [6]), while ERS1/2 images, acquired with λ = 5.66 cm and the same nominal revisiting time as ENVISAT satellite, were processed with the PSInSAR technique (Permanent Scatterers InSAR [7]). Table I summarizes all technical information for each case study.

Table I. Summary of PSI data stacks analyzed for the case studies, with distinction of satellites exploited and related acquisition geometry (ascending = A; descending = D).

Site	Satellite	Geometry	Monitoring period
Capo Colonna	ERS1/2	A	27/03/1995 – 17/10/2000
	ERS1/2	D	18/06/1992 – 17/12/2000
	ENVISAT	A	24/06/2003 – 27/07/2010
Pozzuoli	ERS1/2	D	08/06/1992 – 07/12/2000
	ENVISAT	A	13/11/2002 – 22/10/2008
Agrigento	ERS1/2	D	11/11/1992 – 23/12/2000
	ENVISAT	A	16/11/2002 – 17/07/2010
	ENVISAT	D	05/06/2003 – 18/02/2010

DISCUSSION

Feasibility of PSI measures on cultural heritage

Successful implementation of PSI techniques for analyses on sites with ruins and monuments mainly depends on the availability of sufficient PS density over the study area, detected through the multi-interferogram processing of SAR images. Highly vegetated sites and cultural heritage, partially or totally hidden by wooded coverage, can be monitored hardly, since vegetation prevents the recognition of PS on ground. On the other side, archaeological structures and buildings located in open areas or emerging from forested surroundings can generate PS, useful to carry out deformation analyses. To this purpose, the case of the 17th-century Sanctuary of St. Rosalia in Palermo, Sicily, is quite demonstrative (figure 2A). The building, sacred thanks to the St. Rosalia's relics there preserved, was built within a narrow ravine of the rock mass at the top of Mount Pellegrino. Although the surrounding densely vegetated areas do not provide any PS, they contribute to emphasize the availability of PS in correspondence to the historical structures, as found after the processing of ENVISAT descending data stack (figure 2A). The orientation of the Sanctuary and the satellite acquisition geometry led to the identification of PS in the study area; such sort of evidence confirms the potentials of PSI for applications on archaeological complexes sited in forested areas or wetlands (e.g., Maya temples in Mexico, Angkor monuments in Cambodia).

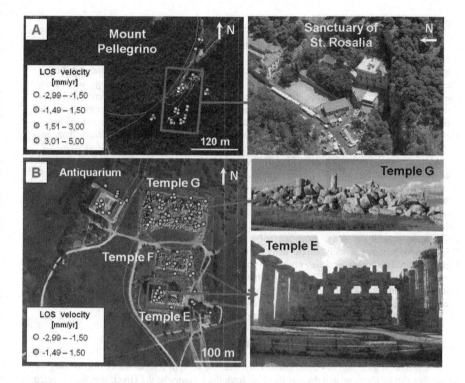

Figure 2. A) PS LOS velocities in the area of St. Rosalia's Sanctuary in Palermo, Sicily, obtained from ENVISAT descending data acquired in November 2002 – October 2010. PS concentrate on the buildings of the Sanctuary. B) PS LOS velocities in the eastern sector of Selinunte, Sicily, obtained from ENVISAT descending data covering the period March 2003 – June 2010. PS are detected in correspondence to both columns drums and standing structures (PSI data in A and B are based on EPRS-E, 2011).

A technical encouragement to test PSI techniques on cultural heritage derives from the typical feature characterizing most archaeological sites, i.e. the presence of ruins and architectural elements (columns, drums, pillars, foundations, wall masonries) emerging from the ground, or even entire structures well preserved in their whole integrity, or reconstructed by archaeologists adopting anastylosis methods. That is the case of archaeological areas like Selinunte, Southwestern Sicily, for which the processing of ENVISAT images resulted in a high density of PS concentrated over the eastern sector of the site, with displacement measures available for both reconstructed (Temple E) and ruined (Temples G and F) structures (figure 2B).

Detection of deformation patterns

One of the most promising perspectives in the use of PSI for deterioration studies is offered by the capability to detect instability processes at 'site scale', allowing the estimation of the magnitude of the observed phenomenon, as well as the characterization of its spatial distribution. The latter parameter plays a fundamental role in the evaluation of the typology of ongoing phenomena, whose intensity can be the same over the entire site, or show a differential behavior. Structural deterioration of archaeological remains is frequently caused by instability of the geologic substratum. The recognition of deformation patterns associable to active natural and/or human-induced processes can clarify the susceptibility of cultural heritage to potential damages, and support a better planning of necessary stabilization interventions.

A deformation pattern is clearly identified whenever the spatial distribution of PS marks an area with homogeneous values of LOS deformation rates and/or the single PS deformation time series display common acceleration/deceleration phases or significant changes in displacement trend. Negative values of LOS deformation rates are generally linked to subsidence/collapse/soil compaction, while positive values are associable to a variety of uplift motions, spanning from volcanic activity to groundwater recharge and clayey soil expansion. In terms of extent, a deformation pattern can range from tens of square meters to hundreds of square kilometers. While *macro-patterns* usually reveal active phenomena at regional scale, *micro-patterns* can highlight conservation criticalities on single portions of a sector or monument, presumably in correspondence to structural weakness points, more deteriorated architectural elements, or structures founded on unstable ground. Non-homogeneous distribution of LOS deformation rates over the interest area can be a reliable evidence of differential response to the same phenomenon, or the coexistence of different phenomena within the same area.

Subsidence and cliff instability in Capo Colonna, Central Calabria

The archaeological site of Capo Colonna occupies the head of a promontory overlooking the Ionian Sea, SE of Crotone, and derives its name from the columns of the Greek temple dedicated to the goddess Hera (figure 3). It formerly constituted the landmark for ancient sailors, but nowadays a unique column still remains on site, after centuries of repeated robbing of building materials. The Column, carved in the local calcarenite and founded on a thick sandstone base, is located at the margin of a composite terrace, very close to the rock cliffs, at about 15 m a.s.l.

The local geologic stratigraphy shows clay bedrock (pelite) outcropping in the southern part of the promontory and constituting the submerged coastal platform. It is overlain by a sequence of calcareous sandstones, with different grade of cementation and intense fracture pattern in the upper part. The constant action of sea erosion undermines the cliff and generates rockfalls and collapses, with a total retreat of 150 m in the last 110 years for the north-eastern tip [8]. Here, the ruins of the ancient Roman settlement are perilously distributed along the cliff edge, directly facing the sea, with high susceptibility to masonry collapses and toppling in consequence of rockfalls. Further geohazard for the preservation of the archaeological remains is represented by the subsidence that historically affects the entire promontory. Scientists recognized the origin of such phenomenon in a combination of natural processes (e.g., lithostratigraphic setting, seismic activity and eustatic sea-level changes) and hydrocarbons extraction [9]. A wide-area worsening of the substratum instability might develop in the near future, as well as local destabilization of the standing structures.

The PSInSAR analysis performed on historical data acquired from the satellites ERS confirms an active subsidence over the whole promontory (figure 3). A huge area characterized by homogeneous deformation pattern with displacements away from the satellite can be identified in the descending data stack in June 1992 – December 2000 (figure 3A). Average LOS velocity of -6 mm/yr and a definite trend away from the satellite are measured since May 1995 for PS distributed within the archaeological site. Going westward inland, LOS velocities tend to increase, reaching values up to -10 mm/yr. Comparison of PS distribution with lithological and marine terraces maps highlights that the subsidence macro-pattern is mainly concentrated in the correspondence to the calcareous terrace, which constitutes the foundation of the archaeological structures. A drastic change in LOS displacement rates is identified W of the terrace. Such evidence suggests a correlation between local geologic setting and the measured displacements.

Subsidence remains active even after 2000, as observed from the PSP-DIFSAR analysis of the ENVISAT ascending data stack (2003-2010) (figure 3B). The spatial distribution of the LOS velocities shows higher values in the central part of the promontory than the coastal area, with values ranging from -5 to -13 mm/yr. Hence, contribution from local groundwater pumping, which overlaps with the effects of local geology, cannot be excluded. As observed in the ERS1/2 descending data (1992-2000), the subsidence pattern follows the extent of the calcareous terrace, with significant values of LOS displacements away from satellite distributed westward inland.

Focusing on the archaeological area, three localized micro-patterns were identified on the following sectors: 1) Column and Temple A; 2) balneum and building M; 3) Capo Nao Tower and St. Mary Sanctuary (figure 3C). Within the sector 1, the Column and the surrounding basement show displacements with average LOS deformation rate of -4.8 mm/yr and two different accelerations recognizable within the PS time series, respectively, in the periods June 2004 – August 2005 and June 2008 – March 2009. Similar acceleration phases characterize PS distributed on the nearby foundations of the Temple A. Localized displacements interest both the ruins and ground at the SW corner of the balneum and building M area (sector 2), with average LOS velocity of -5.6 mm/yr and same changes in displacement trend as those in the PS time series identified on the architectural structures of Capo Nao Tower and the church (sector 3). Movements away from the satellite, with LOS velocity up to -4.3 mm/yr, were detected along the rock cliff. These displacements can be reasonably attributed to a combination of subsidence and rock slope instability. The cliff is extensively fractured in different blocks prone to falling down into the sea, with potential impacts on the structures dangerously located along the cliff edge.

The conservation criticalities discovered for the sectors 1 and 3 agree with risk mapping reported in the Hydrogeological Setting Plan for Capo Colonna promontory, which classifies them as landslide-prone areas. Furthermore, the evidence of subsidence achieved from the radar-interpretation of two data stacks different in exploited satellites, monitoring periods and processing approaches adopted, confirms the persistence of deformation at least since 1992 over the entire archaeological area, and more generally over the promontory, with significant worsening at regional scale, from historical data to more recent measures.

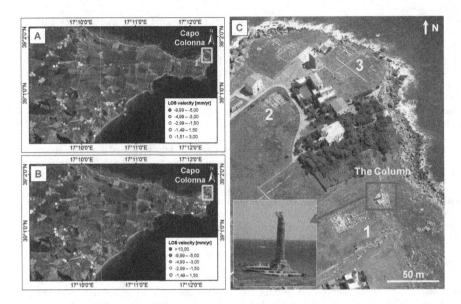

Figure 3. Subsidence macro-pattern detected over Capo Colonna promontory: A) ERS descending data (1992-2000); B) ENVISAT ascending data (2003-2010). Deformation affects the archaeological ruins in the sector: 1. Column and Temple A; 2. balneum and building M; 3. Capo Nao Tower and St. Mary Sanctuary (PSI data in A and B are based on EPRS-E, 2011).

Deformation patterns due to volcanic bradyseism in Pozzuoli, W of Naples

The historic centre of Pozzuoli, formerly the Roman town of *Puteoli*, is located W of Naples, along the coast of the homonymous bay, within the Campi Flegrei area. The latter is an active volcanic caldera, produced during the eruption of the Neapolitan Yellow Tuff (about 15,000 years BP). The entire area is subjected to bradyseism, i.e. gradual uplift/descent vertical movements as effects of filling/emptying of underground magma chamber and/or thermal activity, quite common in volcanic calderas. A long history of local ground elevation changes is recorded, with uplifts having frequently preceded eruptions, as occurred before the event that formed the scoria cone of Monte Nuovo in 1538.

Effects of historical bradyseism are also recognized on the archaeological heritage of Pozzuoli, like the marine burrowing mollusks found up to about 7 m above present sea level on the three marble columns of the Roman marketplace (*Macellum* or *Serapaeum*) and selected by scientists as biological sea level indicators. Analysis of old photographs allowed an estimation of the progressive sinking that interested the columns and the surrounding remains until 1950, with the occurrence of a major uplift in the period 1950-1952 [10]. Attention of local administrators for potential impacts on monuments and public safety is testified by the forced evacuation in 1970 of the entire Rione Terra, the southern historical quarter of Pozzuoli.

Focusing on the last thirty years of bradyseism at Pozzuoli, the Campi Flegrei area was affected by a gradual slow subsidence in 1984-2006, temporarily interrupted by mini-uplift events and modest seismic activity in 1989 and 1999-2000. Nevertheless, recent geodetic measurements suggest a significant inversion of the displacement trend. Uplift movements were measured in November 2004 – November 2006, with a low but increasing rate leading to about 40 cm of uplift till the end of October 2006 [11].

A satellite confirmation for such sequence of differential bradyseismic phases was retrieved from the comparison between ERS descending (1992-2000) and ENVISAT ascending (2002-2008) data stacks processed for the entire area of Campi Flegrei, jointly with a detailed analysis of PS time series identified in the correspondence to archaeological structures (figure 4). ERS data clearly show sub-circular banded subsidence pattern within the Neapolitan Yellow Tuff caldera, centered on Pozzuoli (figure 4A). Here, PS velocities reach the highest values, with average LOS velocity more than -30 mm/yr away from the satellite. PS time series for Serapaeum and Roman amphitheatre display a constant subsidence trend in the period June 1992 – October 1999 with LOS velocity of -27 mm/yr and -29 mm/yr, respectively. An inversion of the trend started in December 1999, with average LOS velocity for the two archaeological monuments of about 31 mm/yr and 36 mm/yr towards the satellite (figure 4B). The presence of similar displacement trends in the PS time series over the Rione Terra defines a deformation pattern spatially extended to the whole town centre, in agreement with geodetic measurements.

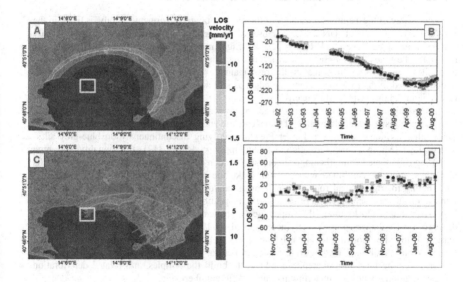

Figure 4. Deformation patterns detected over the Campi Flegrei area and Pozzuoli (the yellow box marks the historic town centre): A) subsidence in ERS descending data (1992-2000); C) uplift in ENVISAT ascending data (2002-2008); with related PS time series (B-D). PSI data are based on EPRS-E (2011).

A confirmation to the uplift phase detected after 2004 is found from the spatial distribution of ENVISAT ascending data (2002-2008) (figure 4C). Ground motions result completely inverted all over the study area, with definite uplift deformation pattern on Pozzuoli and highest values of LOS velocity up to 7 mm/yr towards the satellite for PS identified over the Rione Terra. Uplift also affects the eastern coast of the bay. Differently from the subsidence trend occurred in 1992-1999, PS time series do not show constant uplift trend in 2002-2008. Detailed temporal analysis of PS for Serapaeum, Roman amphitheatre and Rione Terra highlights three major uplift phases, respectively, in the periods November 2002 – August 2003, June 2005 – January 2007, and March 2008 – October 2008 (figure 4D). Such phases were only temporarily interrupted by deceleration phases, with relative stabilization or even the occurrence of displacements away from the satellite. In correspondence to the archaeological monuments, the highest LOS velocities are observed during the second uplift phase, reaching values of 37 mm/yr.

Although both the subsidence and uplift patterns detected in Pozzuoli do not configure differential displacements for the foundation ground of the archaeological monuments, they undoubtedly confirm the critical exposure of the local cultural heritage to sequences of different terrain motions that could induce structural stresses on the masonries, causing progressive destabilization and/or loss of verticality (especially for linear architectural elements like columns and pillars), or even localized collapses.

PS time series analysis on the Valley of the Temples, Agrigento

The Valley of the Temples, located S of the modern town of Agrigento, Sicily, is listed in the World Heritage List of UNESCO because of its archaeological heritage, extraordinarily preserved from the Greek and Roman periods. Eight monumental temples, partially or totally recomposed by the archaeologists, were built between the 6th and 5th centuries BC along the edge of a hill, purposely chosen as natural buttress against external attacks. The defense walls were directly carved in the rock mass along the edge, as still appreciable walking from the Temple of Hera Lacinia to the Temple of Concord.

The geologic setting of the area is characterized by an asymmetrical syncline with E-W axis, whose core consists of an alternation of biocalcarenite and biocalcirudite, marly sand and clayey-sandy silt (Agrigento formation; Lower Pleistocene) lying discontinuously on blue-grey silty-marly clays (Mt. Narbone formation; Middle-Upper Pliocene) [12]. Low quality and high fracturing of biocalcarenite, rheological contrast between calcarenite (brittle) and underlying silts and clays (plastic), combined with erosion, make the entire relief prone to slope instability, contributing to trigger local rock-slumps, topples and/or slides, with high probability of direct impacts on the archaeological heritage. On-site inspections led to discover diffuse presence of large sized fallen blocks along the slope, such as beneath the Temple of Hera Lacinia, where debris of past collapses progressively accumulated. In 1976 a landslide occurred at the eastern margin of the site, blocking temporarily the access to the Valley and damaging the Temple of Hera Lacinia. Additional factors predisposing structural instability have to be identified in the properties of the local biocalcarenite employed to build the temples, and related deterioration processes, among which the undermining due to differential erosion.

The impression that the site is interested by different localized criticalities, rather than a unique instability phenomenon, was confirmed by site-specific analysis of both historical and recent satellite data covering the period 1992-2010. The spatial distribution of LOS deformation rates does not configure any pattern associable to wide-area active phenomena, but highlights single areas with ongoing displacements that might evolve in future damages. In such cases detailed analysis of PS time series represents the most effective approach to distinguish unstable from stable sectors and achieve an up-to-date mapping of conservation criticalities to be verified directly on site [13]. Furthermore, PS that are apparently classifiable as 'stable' due to their low values of average LOS velocity, can actually show a time series characterized by sequence of acceleration/deceleration phases that, during the averaging over the entire monitoring period, results in an overall relative stability. Peculiar behavior of construction materials and terrain can also be observed.

Analyzing ERS descending data (1992-2000), two apparently 'stable' PS, found in correspondence to the architectural elements of the Temple of Concord, show similar displacement trend with a seasonal component and a slight tendency to movement towards the satellite, with average LOS velocity up to 1.3 mm/yr, period of 1 year and amplitude of 5 mm. Something similar is observed for three PS identified on the terrain in front of the eastern façade of the Temple of Heracles, with average LOS velocity up to 1.2 mm/yr (figure 5A). On the other hand, for the PS located SW of the temple the seasonal component of the displacements configures a constant trend away from the satellite, with LOS deformation rate of -5.1 mm/yr.

Movements with opposite direction and average LOS velocity of 2.4 mm/yr are observed in 1992-2000 close to the standing columns of the Temple of Castor and Pollux, within an almost stable area, suggesting the presence of localized instability. A confirmation is found with the ENVISAT ascending data (2002-2010), that detect two PS at the SE corner of the temple, with LOS velocities of -2.8 mm/yr away from the satellite and 1.8 mm/yr towards the satellite, respectively.

Based on the ENVISAT ascending data, some conservation criticalities are also detected for other temples. In the area facing the eastern façade of the Temple of Hera Lacinia, one PS suggests the presence of progressive movements away from the satellite, with average LOS velocity of -1.2 mm/yr in 2002-2010, which may be followed by a further worsening. These movements seem reasonably attributable to local ground instability. On the other hand, the displacements with average LOS velocity of -2.2 mm/yr away from the satellite, measured in correspondence to the standing stone masonry at the pronaos of the Temple of Olympian Zeus, suggest considering them as effects of ongoing structural criticality.

In relation to rock toppling/fall hazard along the cliff edge, a PS with significant deformation trend away from the satellite is found in the ENVISAT descending data stack, in correspondence to boulders highly fractured and toppled along the cliff edge, SW of the Temple of Concord (figure 5B). The average LOS velocity of -4.1 mm/yr calculated in the period June 2003 – September 2008 confirms the high susceptibility of the entire sector, where several toppling events occurred and underwalls were built to prevent further block detachments.

Potentials of PSI techniques to early stage warning of triggering/reactivation of such phenomena is demonstrated by the detection of unexpected movements on an apparently 'stable' sector of the cliff edge, along the walkway towards the Temple of Concord. Although the average LOS velocity is quite low, one PS shows an acceleration phase since August 2008, assumable as precursor signal of a probable block detachment.

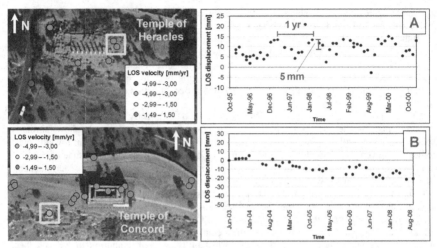

Figure 5. A) PS LOS velocities in the area of the Temple of Heracles, obtained from ERS descending data (1992-2000). Seasonal component with 5 mm of amplitude is detected within the PS time series. B) PS LOS velocities for the Temple of Concord, obtained from ENVISAT ascending data (2003-2010). One PS with constant trend away from the satellite is detected in correspondence to the cliff edge (PSI data in A and B are based on EPRS-E, 2011).

CONCLUSIONS

Satellite interferometric data from the Extraordinary Plan of Environmental Remote Sensing were explored on three cultural heritage sites in Southern Italy to detect ground and structural deformation patterns, and clarify the nature of past/active deterioration phenomena. Site-specific analyses performed on Capo Colonna and Pozzuoli proved the suitability of InSAR techniques to identify wide-area deformation phenomena with direct impact on archaeological monuments. Recognition of localized conservation criticalities over the Valley of the Temples in Agrigento was achieved through detailed analysis of PS time series, discriminating different deformation behaviors and identifying unstable sectors where further investigations and/or interventions might be planned.

ACKNOWLEDGMENTS

PSI data of the Extraordinary Plan of Environmental Remote Sensing (EPRS-E) were made available through the freely accessible WMS service of the National Geoportal of the Italian Ministry of Environment, Territory and Sea (METS).

REFERENCES

1. D. Tapete, N. Casagli, R. Fanti, C. Del Ventisette, R. Cecchi, P. Petrangeli, *Geophysical Research Abstracts* Vol. 13, EGU2011-8387 (2011).
2. N. Casagli, R. Cecchi, D. Tapete, P. Petrangeli, R. Fanti, C. Del Ventisette in *Scienza e Beni Culturali XXVII. 2011 Governare l'innovazione*, G. Biscontin and G. Driussi (eds.) , Edizioni Arcadia Ricerche, Marghera-Venezia, 2011, 323-331.
3. D. Massonnet and K.L. Feigl, *Rev. Geophys.* 36, (1998) 441-500.
4. E. Cabral-Cano, A. Arciniega-Ceballos, O. Díaz-Molina, F. Cigna, B. Osmanoğlu, T. Dixon, C. DeMets, F. Vergara-Huerta, V.H. Garduño-Monroy, J.A. Ávila-Olivera, E. Hernández-Quintero in *Land subsidence, associated hazards and the role of natural resources development*, D. Carreón-Freyre et al. (eds.), Hydrological Sciences Journal, Red Book Series. IAHS Press, Wallingford, UK, 2010, 164-169.
5. P.A. Rosen, S. Hensley, I.R. Joughin, F.K. Li, S.N. Madsen, E. Rodriguez, R.M. Goldstein, *Proc. I.E.E.E.* 88 (2000) 333-382.
6. M. Costantini, A. Iodice, L. Magnapane, L. Pietranera, *Proc IGARSS* (2000) 3225-3227.
7. A. Ferretti, C. Prati, F. Rocca, *IEEE T. Geosci. Remote* 39 (2001) 8-20.
8. U. Chiocchini, in *Geological and Geotechnical influences in the Preservation of Historical and Cultural Heritage*, G. Lollino ed., GNDCI – CNR, n. 2133, Turin, 2000, 389-396.
9. F. Verdecchia, C. Zoccatelli, E. Norelli, R. Miandro, in *Land subsidence, associated hazards and the role of natural resources development*, D. Carreón-Freyre et al. (eds.), Hydrological Sciences Journal, Red Book Series. IAHS Press, Wallingford, UK, 2010, 345-351.
10. C. Del Gaudio, I. Aquino, G.P. Ricciardi, C. Ricco, R. Scandone, *J. Volcanol. Geotherm. Res.* 195 (2010) 48-56.
11. C. Troise, G. De Natale, F. Pingue, F. Obrizzo, P. De Martino, U. Tammaro, E. Boschi, *Geophys. Res. Lett.* 34 (2007) L03301.
12. V. Cotecchia, F. Fiorillo, L. Monterisi, R. Pagliarulo, *Giornale di Geologia Applicata* 1 (2005) 91-101.
13. F. Cigna, C. Del Ventisette, V. Liguori, N. Casagli, *Nat. Hazards Earth Syst. Sci.* 11 (2011) 865-881.

AUTHOR INDEX

SUBJECT INDEX

Printed in the United States
by Baker & Taylor Publisher Services